Formal Languages, Automata and Numeration Systems 2

To Christelle, Aurore and Maxime.

Series Editor
Valérie Berthé

# Formal Languages, Automata and Numeration Systems 2

## Applications to Recognizability and Decidability

Michel Rigo

WILEY

First published 2014 in Great Britain and the United States by ISTE Ltd and John Wiley & Sons, Inc.

ISTE Ltd
27-37 St George's Road
London SW19 4EU
UK

www.iste.co.uk

John Wiley & Sons, Inc.
111 River Street
Hoboken, NJ 07030
USA

www.wiley.com

Library of Congress Control Number: 2014945517

British Library Cataloguing-in-Publication Data
A CIP record for this book is available from the British Library
ISBN 978-1-84821-788-1

# Contents

# Foreword

The interplay between words (in the most general sense, including symbolic dynamics), computability, algebra and arithmetics has now proved its relevance and fruitfulness. Indeed, the cross-fertilization between formal logic and finite automata (such as that initiated by J.R. Büchi), or between logic and aperiodic substitutive tilings (through Wang–Berger–Robinson works) has paved the way to recent dramatic developments. Let us quote, for example, the characterization of the entropies of multidimensional shifts of finite type as right recursively enumerable numbers, by M. Hochman and T. Meyerovitch, or the transcendence results for the real numbers having a "simple" binary expansion, by B. Adamczewski and Y. Bugeaud.

This book is at the heart of this interplay through its unified exposition of the connections between formal languages, automata and numeration, with a number-theoretic flavor. Moreover, objects here are considered with a perspective that comes from both theoretical computer science and mathematics. Here, theoretical computer science offers topics such as decision problems and recognizability issues, whereas mathematics offers concepts such as discrete dynamical systems. Dynamical systems serve here as a common thread and occur in several forms, such as symbolic forms, or as systems of an arithmetic nature (e.g. the Gauss map of continued fractions) in the framework of numeration and arithmetic dynamics, or else, as cellular automata.

But let us come back to the core of this book, namely words. This book provides a systematic treatment of the concepts that have to do with words (and their combinatorics) in all their forms. Here, words are finite or infinite, they are considered alone or they come in sets (languages or subshifts). They can also be multidimensional. They are most often simple in a way that is well explored here through the review of the numerous existing notions of complexity that allow their classification. They are generated by morphisms (these are substitution rules that replace letters by words), they are accepted by simple machines like automata, they are definable with respect to a logical expression, or else, they code classical dynamical systems of an arithmetic nature, such as interval exchanges or Sturmian codings of circle translations.

Words are, moreover, considered from the two complementary viewpoints of word combinatorics and symbolic dynamics. Combinatorics on words studies the combinatorial properties of sequences of symbols (as an illustration, consider the classical subject of occurrences of powers of factors with example the recently solved Dejean's conjecture), whereas symbolic dynamical systems are compact sets of infinite words with dynamics provided by the shift, endowed possibly with a measure which contains probabilistic information on these words (e.g. the frequencies of their factors). All these approaches yield different questions, classifications, etc.

But why are words and their study so important? It is because of their representation power: they are a natural way to code elements of an infinite set using a finite amount of symbols. In this context, numeration systems play a specific role. And their treatment in this book is again a perfect illustration of the interconnection between computer science and mathematics. As an example, an arithmetic property of the base of a numeration system, such as being a Pisot number, is revisited from the viewpoint of recognizability, by exploring the way arithmetic properties are reflected in the representation. Similarly, finite automata are known to have strong descriptive power. This is natural if we think of finite automata as representations of finitely generated subgroups of a free group. This has also induced, in the direction of computational and geometric group theory, the fascinating development of automatic groups. But finite automata also serve, of course, as language acceptors, with logic playing an increasing role for the study of regular languages; just think of the recent development of the theory of regular cost functions which is a quantitative extension of the classical notion of regularity. It is in this context that one of the most original chapters of this book (at least in my view) takes place, devoted to the use of formal methods applied to morphic words, through automatic certification of properties expressed as a first-order formula in a suitable system (see Chapter 3).

A few words now concerning the "genesis" of this book. I have known Michel for many years and I have seen him develop into an impressive lecturer, famous for his work in popularizing mathematics, in addition to also being a highly recognized expert in his field. When I suggested that he use his teaching experience to write about numeration and automata, he responded enthusiastically. The result is this book written in Michel's personal style and with his sense of humor. It will serve as a clear and accessible introduction to the field, being a highly valuable tool for teaching and making the subject accessible to a wide audience, thanks to its numerous exercises, self-contained chapters and very large number of examples. But, by being connected to the most recent research developments, it will also serve as a reference book to experts in the area who will still learn new things. I am surprised both by the extent of its coverage and by the impressive variety of examples issued from the various viewpoints. I hope you enjoy reading it as much as I did.

Valérie Berthé
July 2014

# Introduction

This book, comprised of two volumes, is a somewhat extended version of lectures basically dedicated to combinatorics on words and numeration systems that I am giving at the University of Liège. The course is usually (but not necessarily) followed by students interested in discrete mathematics or theoretical computer science. The chosen level of abstraction should allow undergraduate students to study the exposed topics.

## I.1. What this book is or is not about

In the long process of writing this book, I have expanded my initial notes with many examples and many extra concepts to acquire a consistent overview of the field. Nevertheless, this book is *not* intended to serve as an encyclopedic reference.

I have picked some of my favorite topics in the area and I have also decided to shorten the presentation of some items (not because there are less interesting but choices had to be made to keep this book reasonably short). Indeed, the book most probably reflects what I myself prefer: I am always more interested in the *combinatorics* and the underlying *discrete structures* arising from a problem.

When preparing this book, I chose to present a fairly large variety of basic notions and important tools and results. Sometimes, I only give an overview of a subject and *proofs are, therefore, omitted*. For the reader wanting to study a specific topic further, many pointers to the relevant bibliography are given and each chapter ends with notes and comments. Indeed, the main goal of this book is to give *quick access to actual research topics at the intersection between automata and formal language theory, number theory and combinatorics on words*.

## I.2. A few words about what you will find

The notion of a word, i.e. a (finite or infinite) sequence of symbols belonging to a finite set, is central throughout this book. It has connections with many branches of mathematics and computer science: number theory, combinatorics, formal language theory, mathematical logic, symbolic dynamics, coding theory, computational complexity, discrete geometry, stringology, etc.

*Combinatorics on words*. We can be interested in the combinatorial properties of finite or infinite sequences of symbols over a finite alphabet: what the possible arrangements are, how many such configurations can be achieved and so on. As a trivial example, over a binary alphabet any word of a length of at least 4 contains a repeated factor of the kind $uu$ (try to prove it). Therefore, we can look at patterns that are unavoidable in sufficiently long sequences or count the number of patterns or configurations that may appear in a particular context. These are some of the general questions that will be considered in Volume 1, [RIG 14]. In particular, we will concentrate on infinite words that can be obtained by a simple procedure consisting of the iteration of a morphism over a free monoid. We will mostly deal with a large class of self-similar words: the so-called morphic words and, in particular, and with automatic sequences that are generated by a constant-length morphism.

*Formal language theory.* A language is merely a set of words. In this book, we will mostly encounter languages of finite words. One exception is a short incursion into symbolic dynamical systems with the language of the $\beta$-expansions of the real numbers in the interval $[0, 1)$. Chomsky's hierarchy introduced in the theory of formal languages provides a classification depending on the machine needed to recognize an infinite language of finite words. From a computational perspective, the simplest languages are the regular languages. They are accepted (or recognized) by finite automata, and described by regular expressions. Chapter 1, is a short chapter presenting the main properties of these languages. We will constantly see connections existing between regular languages, automatic sequences and numeration systems. For instance, quite often we associate a finite automaton with a morphism.

*Number theory.* A finite word can also be used to represent an integer in a given numeration system (e.g. integer base expansions and many other non-standard systems are discussed in depth in several chapters of this book). To quote A. Fraenkel: "There are many ways of representing an integer uniquely!" [FRA 85]. Similarly, an infinite word can represent a real number or the characteristic sequence of a set of integers. With that respect, a natural question is to study links existing between arithmetical properties of numbers (or sets of numbers) and syntactical properties of their expansions. Chapter 2, is dedicated to numeration systems with a particular emphasis on words representing numbers. Indeed, the chosen numeration system has a strong influence on the syntactical properties of the corresponding representations. A cornerstone is the notion of a recognizable set of numbers whose elements, when represented within a given numeration system, are recognized by a finite automaton.

*Formal methods applied to infinite words and sets of numbers.* In Chapter 3 of this Volume, I describe a recent trend in combinatorics on words. Due to automata theory and Büchi's theorem, we will see how formal methods enter the frame regarding decision problems, or automatic theorem-proving, relevant in combinatorics on words. If a property about some infinite words can be described by a well-written logical formula, then this property can be tested automatically. Such a procedure holds for a large class of infinite words generated by iterated morphisms (for automatic sequences and those stemming from Pisot numeration systems as presented in this book). The expressiveness of Presburger arithmetic (with an extra predicate) provides an interesting alternative to dealing with a sufficiently large class of problems about infinite morphic words. We can imagine automated certificates for several families of combinatorial properties. But the price to pay is that we would have to deal with fairly large automata. It is a field of research where combinatorists and computer scientists can work together fruitfully: on the one hand, it is well-known that, in the worst-case, the obtained decision procedures can be super-exponential, but on the other hand, the considered problems about words seem to be of relatively small complexity.

## I.3. How to read this book

The goal is that, after reading this book (or at least parts of this book), the reader should be able to fruitfully attend a conference or a seminar in the field. I hope that the many examples presented along the text will help the reader to get some feeling about the presented topics even though we are not going too far in the technical aspects of the proofs. Also, prerequisites are minimal. We will not explore topics requiring measure theory or advanced linear algebra (we have avoided results related to Jordan normal form of matrices) or non-elementary number theory. Two sections are devoted to results in algebraic number theory and formal series. Sections 1.1.2 and 1.2.2 of Volume 1 serve as references that the reader may consult when needed. Sections 3.1 and 3.2 give a self-contained presentation of the concepts of mathematical logic needed in this book. Those rigorous and technical sections should not discourage the reader to pursue his/her study. Most of the material can be accessed without much background.

My initial aim was to quickly get to the point but it seems that the stories I wanted to tell were indeed quite longer than I initially thought. I have to confess that writing this book was a quite unexpected adventure (I was perpetually trying to meet the deadlines and also dealing with my other duties at the University and at home).

There are several paths that the reader can follow through this book. Some are quite long, some are shorter.

– For a *basic introduction*, I propose reading parts of Chapter 1, Volume 1 (skipping the reference sections), Chapter 2, again Volume 1, up to and including section 2.4. If the reader already has some knowledge about automata, then we can conclude with Chapter 3 of this volume, concentrating on results about integer base systems.

– For a one-semester course in *combinatorics on words*, I propose a reading of Volume 1, not sacrificing the rigorous presentation of section 1.2.1, Volume 1.

– For a *numeration system* oriented reading, again organized over one semester: browse through the first chapter (with a careful reading of the examples related to numeration systems), then go to section 2.3, of Volume 1, and conclude with the last two chapters of this volume.

– For a course oriented toward *interaction between automata, logic and numeration systems*, we can focus on Chapters 1 and 3 of this volume.

About other sources treating similar subjects, an excellent companion for this book is definitely *Automatic Sequences: Theory, Applications, Generalizations* [ALL 03a] written by Allouche and Shallit. I do hope that the two books can be read independently and can benefit from each other. There is also a non-zero intersection with several chapters of the Lothaire's book *Algebraic Combinatorics on Words* (namely those about Sturmian words written by Berstel and Séébold and the one on numeration systems written by Frougny) [LOT 02]. Some chapters of the volume *Combinatorics, Automata and Number Theory* [BER 10] as well as [PYT 02] can also serve as a follow up for the present book. In particular, Cassaigne and Nicolas's chapter on factor complexity is a natural continuation for our Chapter 2, Volume 1. I should finish by mentioning two papers that were very influential in my work: [BRU 94] and [BRU 95]. With this book, I hope that the reader would learn as much material as found in these two papers.

Tags of bibliographic entries are based on the first three letters of the last name of the first author and then the year of publication. In the bibliography, entries are sorted in alphabetical order using these tags.

I intend to make a page:

`www.discmath.ulg.ac.be/flans/`

for errata and comments about this book.

## I.4. Acknowledgments

I would like to express my gratitude to Valérie Berthé for her constant and enthusiastic support, for the many projects we run together and finally, for her valuable comments on a draft of this book.

Several researchers have spent some precious time reading a first draft of this book, their careful reading, their feedback and expert comments were very useful and valuable: Anna Frid, Julien Leroy, Aline Parreau, Narad Rampersad, Eric Rowland, Aleksi Saarela and Jeffrey Shallit. They proposed many clever improvements of the text. I warmly thank them all. I would like to give a special thank to Véronique Bruyère for comments on the last chapter.

I also sincerely thank Jean-Paul Allouche, Émilie Charlier, Fabien Durand and Victor Marsault for their feedback.

Even though he was not directly involved in the writing process of this book, the first half of the book has greatly benefited from the many discussions I had with Pavel Salimov when he was a postdoctoral fellow in Liège. Naturally, all the discussions and interactions I could have had along the years with students, colleagues and researchers worldwide had some great influence on me (but such a list would be too long) and I thank them all.

Michel RIGO
July 2014

# 1

# Crash Course on Regular Languages

The theory of finite automata has preserved from its origins a great diversity of aspects. From one point of view, it is a branch of mathematics connected with the algebraic theory of semigroups and associative algebras. From another point of view, it is a branch of algorithm design concerned with string manipulation and sequence processing. It is perhaps this diversity that has enriched the field to make it presently one with both interesting applications and significant mathematical problems.

Dominique Perrin [PER 90]

This chapter is intended to be short. Automata theory and formal language theory have been developed for more than 50 years. There are many textbooks devoted to these theories (and we can easily find series of exercises). To cite just a few (all these books cover in particular what is nowadays considered as classical material): [HOP 79] or [SUD 06] where the focus is oriented toward the computational models and the corresponding algorithms, the comprehensive [SAK 09] where a wider perspective, a more general algebraic framework and emphasis on the underlying structures are provided. I personally like the presentation and the material covered in [SHA 08]. I can also mention [LAW 04] (for a light introduction to the syntactic monoid of a language) or the classic [EIL 74]. See also, the survey [YU 97] or [PER 90] for a condensed exposition.

We have already encountered the notion of automaton in several sections of this book: the formal presentation of deterministic automaton first appeared in definition 2.23. Then the extended model with output was given in definition 2.27, Volume 1, to deal with automatic sequences. We will work only with these types of finite machines[1].

First, we will provide summary of some basic results about finite automata and regular languages. Then, we will select some particular topics related to the main themes of the book: recognizable sets of numbers and morphic or automatic words. The lastsection will present regular languages of polynomial growth. Thus, their characterization is applied to growing letters in a morphic word. Also, this chapter will serve as a preparation for the last chapter of the book where automata will play an important role when dealing with decision procedures. For an easy understanding of Chapter 3, the readers familiar with automata theory should consult sections 1.3 and 1.7.

## 1.1. Automata and regular languages

We begin with a definition that we have already described in definition 2.23, Volume 1. But, here the focus is put on the words, and thus on the language that is accepted by such a machine.

DEFINITION 1.1.– A *deterministic finite automaton*, or DFA for short, over an alphabet $B$ is given by a 5-tuple $\mathcal{A} = (Q, q_0, B, \delta, F)$, where $Q$ is a finite set of states, $q_0 \in Q$ is

---

1 The readers could also remember that cellular automata were introduced in section 1.3.1, Volume 1, but such a model of computation is out of the scope of the present chapter.

the initial state, $\delta : Q \times B \to Q$ is the transition function and $F \subseteq Q$ is the set of final states. The map $\delta$ can be extended to $Q \times B^*$ by setting $\delta(q, \varepsilon) = q$ and $\delta(q, wa) = \delta(\delta(q, w), a)$ for all $q \in Q$, $a \in B$ and $w \in B^*$. The language *accepted* or *recognized* by $\mathcal{A}$ is

$$L(\mathcal{A}) = \{w \in B^* \mid \delta(q_0, w) \in F\}\,.$$

A *regular language* is a language accepted by some DFA.

Given a DFA, we have a partition of $B^* = L(\mathcal{A}) \cup (B^* \setminus L(\mathcal{A}))$. A word $w$ is trivially either accepted whenever $\delta(q_0, w) \in F$ or, rejected whenever $\delta(q_0, w) \notin F$. Since the automaton is deterministic[2], exactly one of the two situations occurs. So to speak, given a word $w$, we start reading this word from the initial state, one letter at a time from left to right. Following the transitions of $\delta$, the current state is updated until the whole word has been read. At that time, we test if the reached state is final or not.

EXAMPLE 1.2.– Consider the DFA over $\{a, b\}$ having $\{0, 1, 2\}$ as set of states. The transition function is defined by $\delta(0, a) = 0$, $\delta(0, b) = 1$, $\delta(1, a) = 2$, $\delta(1, b) = 1$, $\delta(2, a) = 2$, $\delta(2, b) = 2$. The initial state (represented with an in-coming arrow) is 0, and the final states are 0, 1 (represented with an out-going arrow). This automaton is depicted in Figure 1.1. For a word $w = w_0 \cdots w_\ell$, we can consider the behavior (or state transition sequence) of the DFA, that is the sequence of $\ell + 2$ states

$$q_0, \delta(q_0, w_0), \ldots, \delta(q_0, w_0 \cdots w_i), \ldots, \delta(q_0, w_0 \cdots w_\ell)\,.$$

---

2 Also, the domain of $\delta$ is the whole set $Q \times B^*$. If we allow $\delta$ to be a partial function, i.e. defined on a subset of $Q \times B^*$, then $\delta(q, w)$ could be undefined for some $w \in B^*$, in that case, the word $w$ is rejected. It is indeed common to consider only states that may lead eventually to some final state. Therefore, if we remove useless states (and the corresponding transitions), this means that the corresponding transition function is a partial function. In such a case, the automaton is still considered to be deterministic even though *stricto sensu* the domain of $\delta$ should be the whole set $Q \times B$ and by extension $Q \times B^*$.

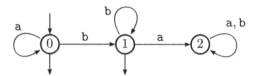

**Figure 1.1.** *A DFA accepting* $\{a^i b^j \mid i, j \geq 0\}$

Reading aabb or ababb gives, respectively,

$$0 \xrightarrow{a} 0 \xrightarrow{a} 0 \xrightarrow{b} 1 \xrightarrow{b} 1$$

$$\text{and} \quad 0 \xrightarrow{a} 0 \xrightarrow{b} 1 \xrightarrow{a} 2 \xrightarrow{b} 2 \xrightarrow{b} 2 \,.$$

The word aabb is accepted because $\delta(q_0, \text{aabb}) = 1 \in F$ but ababb is not because $\delta(q_0, \text{ababb}) = 2 \notin F$. It is easy to see that the language accepted by this DFA is $\{a^i b^j \mid i, j \geq 0\}$.

As we will see soon (when dealing with operations on regular languages), it is also interesting to introduce a more general model of machines: the non-deterministic automata.

DEFINITION 1.3.– A *non-deterministic finite automaton*, or NFA for short, over an alphabet $B$ is given by a 5-tuple $\mathcal{N} = (Q, I, B, \Delta, F)$, where $Q$ is a finite set of states, $I \subseteq Q$ is the set of initial states, $\Delta \subset Q \times B^* \times Q$ is the (finite) transition relation and $F \subseteq Q$ is the set of final states. A word $w$ is *accepted* by $\mathcal{N}$ if there exists an integer $i$, some (possibly empty) words $v_1, \ldots, v_i$ and a sequence of states $q_1, \ldots, q_{i+1}$ such that $w = v_1 \cdots v_i$ and

– $(q_1, v_1, q_2), (q_2, v_2, q_3), \ldots, (q_i, v_i, q_{i+1}) \in \Delta$
– $q_1 \in I$, $q_{i+1} \in F$.

Otherwise stated, there is at least one accepting path from an initial state to some final state with label $w$. The *language accepted* by $\mathcal{N}$ is the set of accepted words. We can assume that $(q, \varepsilon, q)$ belongs to $\Delta$ for all $q \in Q$.

The acceptance condition for an NFA $\mathcal{N}$ can be written in a different way. It is useful to introduce the following notation (that we will use in the determinization algorithm). Let $R \subseteq Q$ be a subset of states of $\mathcal{N}$. Let $w$ be a finite word. We let

$$R.w$$

to denote the set of states defined as follows. The state $r \in Q$ belongs to $R.w$ if and only if there exists an integer $i$, some (possibly empty) words $v_1, \ldots, v_i$ and a sequence of states $q_1, \ldots, q_i$ such that $w = v_1 \cdots v_i$,

$$q_1 \in R \quad \text{and} \quad (q_1, v_1, q_2), (q_2, v_2, q_3), \ldots, (q_i, v_i, r) \in \Delta .$$

Otherwise stated, there is a path of label $w$ starting in a state belonging to $R$ and leading to $r$. Therefore, the language accepted by $\mathcal{N}$ is

$$\{w \in B^* \mid I.w \cap F \neq \emptyset\} .$$

EXAMPLE 1.4.– We have depicted an NFA in Figure 1.2. The readers may notice that we have two initial states: 0 and 3. There are several $\varepsilon$-transitions, i.e. transitions labeled by the empty word, for instance, from 1 to 4. Reading the symbol a from state 0 can lead to any of the four states $0, 1, 3, 4$. With the notation introduced above

$$\{0\}.a = \{0, 1, 3, 4\} = \{0, 3\}.a$$

and $\{5\}.b = \{1, 2, 4, 5\}$. Indeed we have, for instance, that

$$(5, \varepsilon, 1), (1, b, 1), (1, \varepsilon, 4) \in \Delta .$$

There are several ways to accept the word aa (but also some non-accepting paths, although the existence of one accepting path is enough).

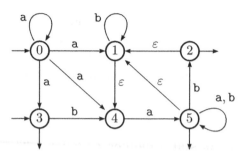

**Figure 1.2.** *A non-deterministic automaton over* $\{a, b\}$

REMARK 1.5.– Every DFA is a special case of an NFA. It has a unique initial state, and the relation $\Delta$ is a subset of $Q \times B \times Q$ that is the graph of a function.

We might suspect that the set of languages accepted by deterministic finite automata over $B$ is a (strict) subset of the set of languages accepted by non-deterministic automata over $B$. Nevertheless, any language accepted by an NFA is also accepted by a DFA (this is the Rabin–Scott theorem from 1959). We do not prove this result here, but below we present a determinization algorithm (without showing its correctness).

First, we can assume that the given NFA is *elementary*. This means that every transition $(q, w, q') \in \Delta$ is such that $|w| \leq 1$. Indeed, if we have a transition $(q, w, q')$, where $w = w_1 \cdots w_\ell$ is a word of length $\ell \geq 2$, then we can replace this transition with $\ell$ new transitions (where the states $q_1, \ldots, q_{\ell-1}$ are newly created states):

$$(q, w_1, q_1), (q_1, w_2, q_2), \ldots, (q_{\ell-2}, w_{\ell-1}, q_{\ell-1}), (q_{\ell-1}, w_\ell, q')$$

For instance, the NFA in Figure 1.2 is elementary.

ALGORITHM 1.6 (subsets construction).– The input is a non-deterministic automaton $\mathcal{N} = (Q, I, B, \Delta, F)$ accepting some language $L$. We can assume that $\mathcal{N}$ is elementary. The output is a DFA $\mathcal{A}$ accepting the same language $L$. The states of $\mathcal{A}$ are subsets of $Q$. The deterministic automaton $\mathcal{A}$ is obtained as follows.

– The initial state of $\mathcal{A}$ is $I.\varepsilon \subseteq Q$. It is the unique state of $\mathcal{A}$ produced when initializing the algorithm.

– At each step of the algorithm, for each state $S$ in $\mathcal{A}$ newly created during the previous step, compute $S.b$ for all $b \in B$. These are states of $\mathcal{A}$, i.e. subsets of states of $\mathcal{N}$. If some new state appears, then it is a newly created state (and we iterate this step of the procedure).

The algorithm stops when no new state is created (the procedure always terminates because there are at most $2^{\mathrm{Card}(Q)}$ states in $\mathcal{A}$).

– The transition function of $\mathcal{A}$ is $\delta(S,b) = S.b$ for all states $S$ and $b \in B$.

– A state $S$ of $\mathcal{A}$ is final whenever $S \cap F \neq \emptyset$.

Because the algorithm takes into account only states created from some previous step, the DFA $\mathcal{A}$ computed by the above algorithm is an *accessible* automaton, i.e. for every state $q$ in $\mathcal{A}$, there exists a word $w$ such that reading $w$ from the initial state of $\mathcal{A}$ leads to $q$. Otherwise stated, every state can be reached from the initial state.

We can also see that if an NFA has $n$ states, we could possibly obtain some DFA with up to $2^n$ states. In some cases, such an exponential blow-up is unavoidable[3]. Such an observation is crucial when dealing with practical

---

3 Let $p_1,\ldots,p_k$ be the first $k$ prime numbers. Consider the language $L_k = \{a^n \mid n \in p_1\mathbb{N} \cup \cdots \cup p_k\mathbb{N}\}$. It is not difficult to prove that this language is accepted by an NFA with $p_1 + \cdots + p_k$ states, but any DFA accepting $L_k$ has at least $p_1 \cdots p_k$ states. (Left as an exercise.) Working over a unary alphabet reduces the arguments to number-theoretic results like the application of Bézout's identity.

implementations like those that we will consider in Chapter 3.

EXAMPLE 1.7.– If we apply the subsets construction to the NFA given in Figure 1.2, we get the following subsets as states of a DFA

$$\{0,3\}, \{0,1,3,4\}, \{4\}, \{0,1,3,4,5\}, \{1,4\}, \{1,4,5\}, \emptyset, \{1,2,4,5\}.$$

Starting with $\{0,3\}$, we compute sets of the form $R.a$ and $R.b$ as follows.

| $R$ | $R.a$ | $R.b$ |
|---|---|---|
| $\{0,3\}$ | $\{0,1,3,4\}$ | $\{4\}$ |
| $\{0,1,3,4\}$ | $\{0,1,3,4,5\}$ | $\{1,4\}$ |
| $\{4\}$ | $\{1,4,5\}$ | $\emptyset$ |
| $\{0,1,3,4,5\}$ | $\{0,1,3,4,5\}$ | $\{1,2,4,5\}$ |
| $\{1,4\}$ | $\{1,4,5\}$ | $\{1,4\}$ |
| $\{1,4,5\}$ | $\{1,4,5\}$ | $\{1,2,4,5\}$ |
| $\emptyset$ | $\emptyset$ | $\emptyset$ |
| $\{1,2,4,5\}$ | $\{1,4,5\}$ | $\{1,2,4,5\}$ |

The final states are the subsets containing a final state of the initial NFA:

$$\{0,3\}, \{0,1,3,4\}, \{0,1,3,4,5\}, \{1,4,5\}, \{1,2,4,5\}.$$

The corresponding DFA is depicted in Figure 1.3.

THEOREM 1.8.– The set of regular languages over a fixed alphabet $B$, i.e. languages accepted by some finite (deterministic or non-deterministic) automaton, is closed under the following operations: union, complement, concatenation, Kleene star, image by morphism and reversal. Moreover, all the corresponding automata can be effectively obtained.

PROOF.– Let $\mathcal{A} = (Q, q_0, B, \delta, F)$ and $\mathcal{A}' = (Q', q_0', B, \delta', F')$ be two DFAs. We build new automata based on $\mathcal{A}$ and $\mathcal{A}'$.

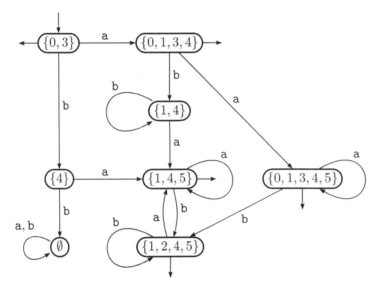

**Figure 1.3.** *Determinization of a non-deterministic automaton over* $\{a, b\}$

– The language $L(\mathcal{A}) \cup L(\mathcal{A}')$ is accepted by the NFA obtained by "merging" the automata $\mathcal{A}$ and $\mathcal{A}'$ (there is no transition between the two automata). We have one "big" automaton that is the disjoint union of the first two ones. It simply has two initial states $q_0$ and $q'_0$. The set of states is $Q \cup Q'$. Indeed, a word is accepted if it is accepted by at least one of the two automata.

– The language $B^* \setminus L(\mathcal{A})$ is accepted by the DFA obtained from $\mathcal{A}$ where the updated set of final states is $Q \setminus F$.

– The language $L(\mathcal{A})\, L(\mathcal{A}')$ is accepted by the NFA obtained by considering sequentially the two automata: the unique initial state is $q_0$, the final states are those of $\mathcal{A}'$ (the states in $F$ are no longer final), and we add $\varepsilon$-transitions from all states in $F$ to $q'_0$.

– The language $L(\mathcal{A})^*$ is accepted by the NFA where we add a new state $r_0$ to $\mathcal{A}$ and $\varepsilon$-transitions from $r_0$ to $q_0$ and from every state in $F$ to $r_0$. The state $r_0$ is the unique initial and final state of this new automaton.

– Let $f : B^* \to A^*$ be a morphism. We consider the NFA where each label $b$ is replaced accordingly with $f(b)$. It is indeed an NFA because the two letters can have the same image by $f$ or the morphism can be erasing. This NFA accepts the language $f(L(\mathcal{A}))$.

– For the reversal of $L(\mathcal{A})$, consider the NFA where the set of initial states is $F$, the unique final state is $q_0$ and all the transitions have been reversed: if $\delta(q, b) = r$ then $(r, b, q)$ belongs to the transition relation of the NFA.    ∎

In many constructions (like the conjunction of two logical formulae discussed in the last chapter), the intersection of two languages is an important operation to carry on. Slightly modifying the construction given below also allows us to derive – almost for free – a closure result about the shuffle of two regular languages (even though such an operation is not as useful as an intersection[4]).

THEOREM 1.9.– The set of regular languages over a fixed alphabet $B$, i.e. languages accepted by some finite (deterministic or non-deterministic) automaton, is closed under intersection and shuffle.

To show that the set of regular languages is closed under intersection and shuffle, we introduce the *product* of two DFAs. The idea is that the product should mimic (i.e. keep track of) the behavior of both original automata.

DEFINITION 1.10.– Let $\mathcal{A} = (Q, q_0, B, \delta, F)$ and $\mathcal{A}' = (Q', q_0', B, \delta', F')$ be two DFAs over $B$. Consider the *product*

---

4 Nevertheless, an interesting application is to compare the growth function (definition 1.33, Volume 1) of the languages $L$, $M$ and $L \sqcup\!\sqcup M$. See, for instance, [RIG 02a] where the shuffle product is used to obtain languages with a given polynomial growth function.

*automaton* $\mathcal{P}$ where

   – $Q \times Q'$ is the set of states;

   – $(q_0, q_0')$ is the initial sate;

   – The transition function $\lambda : (Q \times Q') \times B \to Q \times Q'$ is defined by

$$\lambda((q, q'), b) = (\delta(q, b), \delta'(q', b)), \quad \forall (q, q') \in Q \times Q',\ b \in B.$$

In the product automaton, if the set of final states is $F \times F'$, then this automaton accepts exactly $L(\mathcal{A}) \cap L(\mathcal{A}')$.

EXAMPLE 1.11.– In Figure 1.4, at the top we have presented a DFA accepting the language $\{a^i b^j \mid i, j \geq 0\}$ considered in example 1.2. On the lower left, we have presented a DFA accepting words over $\{a, b\}$ with no factor bb. On the lower right of the figure, we have depicted the accessible part of the corresponding product automaton.

Recall that the (ordinary) shuffle of two finite words $u$ and $v$ is the set $u \sqcup\!\sqcup v$ of words obtained when merging $u$ and $v$ from left to right, but choosing the next symbol arbitrarily from $u$ or $v$. As an example (having words over disjoint alphabets permits us to easily compute the shuffle),

$$ab \sqcup\!\sqcup cd = \{abcd, acbd, acdb, cabd, cadb, cdab\}.$$

We can define accordingly the shuffle of two languages $L$ and $M$ as

$$L \sqcup\!\sqcup M = \bigcup_{u \in L, v \in M} u \sqcup\!\sqcup v.$$

We can easily modify the product automaton given in definition 1.10 to accept the language $L(\mathcal{A}) \sqcup\!\sqcup L(\mathcal{A}')$ where $\mathcal{A}$ and $\mathcal{A}'$ are DFAs. First, we can assume that $\mathcal{A}$ and $\mathcal{A}'$ are defined over disjoint alphabets $B$ and $B'$ (if this is not the case, we can replace one of the two alphabets with new symbols, and we apply an injective morphism $h : B^* \to B'^*$ that maps $b_i$ to $c_i$). Consider the same construction as in definition 1.10 but consider the transition function defined by

$$\lambda'((q, q'), c) = \begin{cases} (\delta(q, c), q'), & \text{if } c \in B; \\ (q, \delta'(q', c)), & \text{if } c \in B'. \end{cases}$$

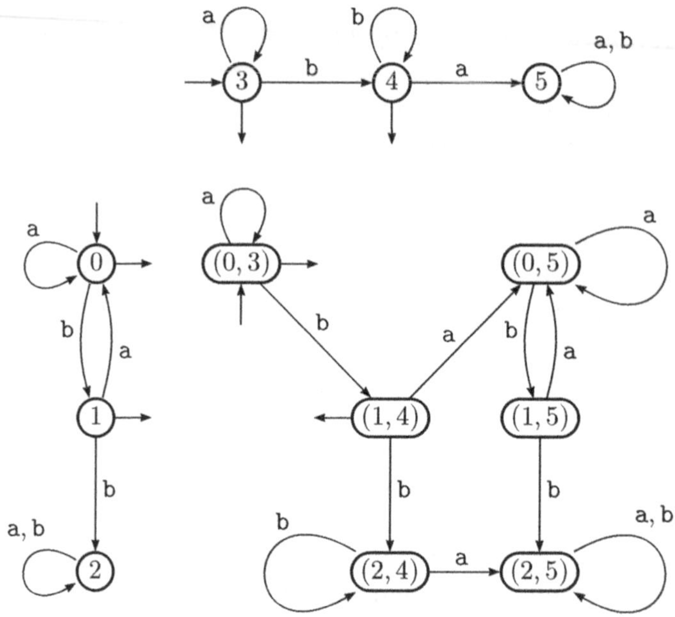

**Figure 1.4.** *Two DFAs and the corresponding product*

Note that $\lambda'$ is well-defined because $B \cap B' = \emptyset$. The set of final states is, as for the intersection of languages, $F \times F'$. Having two disjoint alphabets permits us to act precisely on one of the two automata. If initially, the two automata were not given over disjoint alphabets, we can finish the proof by applying the morphism $h^{-1}$ to the labels belonging to $B'$.

REMARK 1.12.– Every finite language is regular. The readers will be convinced by the following example. Take the language {abba, ab, bba}. Simply consider the subgraph of the full binary tree as depicted in Figure 1.5. In particular, this means that adding or removing a finite number of words to a regular language will not alter its regularity. Regular languages are closed under finite modifications.

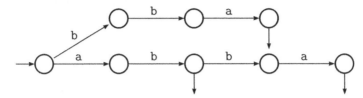

**Figure 1.5.** *A DFA accepting* {abba, ab, bba}

To conclude with this introductory section, we say a few words about regular expressions. DFAs and NFAs are acceptors of regular languages, while regular expressions are generators of these languages. Let $B$ be an alphabet not containing the symbols $+, \cdot, *, (,), e, 0$. A *regular expression* over $B$ is a word over $B \cup \{+, \cdot, *, (,), e, 0\}$ that is obtained by applying a finite number of times the following rules:

– $0$, $e$ and $b$, for all $b \in B$ are regular expressions;

– if $R, S$ are regular expressions, so are $(R + S)$, $(R \cdot S)$ and $R^*$.

For instance, $(((a \cdot b)^* + e) + ((a \cdot b)^* + e)^*)$ is a regular expression. We have just defined a way to produce syntactically "valid" expressions. Now we add some semantics: with each regular expression a language is associated. Let $\mathrm{Reg}(B)$ denote the set of regular expressions over $B$. We define a map $\varphi : \mathrm{Reg}(B) \to B^*$ as follows. Let $R, S \in \mathrm{Reg}(B)$. We set

– $\varphi(0) = \emptyset$;

– $\varphi(e) = \{\varepsilon\}$;

– $\varphi(b) = \{b\}$ for all $b \in B$;

– $\varphi((R + S)) = \varphi(R) \cup \varphi(S)$;

– $\varphi((R \cdot S)) = \varphi(R)\, \varphi(S)$;

– $\varphi(R^*) = (\varphi(R))^*$.

It is common to identify a regular expression $R$ with the language $\varphi(R)$. Also, several expressions may correspond to the same language. Kleene's theorem states that a language is regular if and only if it is generated by some regular expression. There are several classical methods for building an automaton accepting the language described by a regular expression. Conversely, from a regular expression $R$, we can derive an automaton that accepts $\varphi(R)$. From this, the readers may be convinced that the set of regular languages over $B$ is the smallest collection of languages containing the finite languages and closed under union, concatenation and Kleene's star.

## 1.2. Adjacency matrix

The adjacency matrix of a DFA $\mathcal{A} = (Q, q_0, B, \delta, F)$ is a matrix $\mathsf{M}(\mathcal{A})$ whose entries are indexed by $Q$ and where $(\mathsf{M}(\mathcal{A}))_{p,q} = \mathrm{Card}\{b \in B \mid \delta(p, b) = q\}$.

Note that this matrix is similar to the one introduced in section 2.2.1, Volume 1, and associated with a morphism. Actually, take a morphism $f$ and the associated matrix $\mathsf{M}_f$. We have associated with some automaton $\mathcal{A}_f$; see, for instance, examples 2.24 and 2.31, Volume 1. If $f$ is a constant-length morphism, then the DFA $\mathcal{A}_f$ has a total transition function. Otherwise, the transition function is partial. More about these links is discussed at the end of section 1.8.

It is easy to see that the adjacency matrix of the DFA $\mathcal{A}_f$ is the transpose of the matrix $\mathsf{M}_f$. We could decide to have a uniform presentation, but if we stick to the usual conventions, then let us make a distinction when dealing with morphisms or automata. In any event, spectral properties of matrices are invariant under transposition.

PROPOSITION 1.13.– Let $\mathcal{A} = (Q, q_0, B, \delta, F)$ be a DFA. Let $\mathsf{M}(\mathcal{A}) \in \mathbb{N}^{Q \times Q}$ be the adjacency matrix of $\mathcal{A}$. The number of paths of length $n \geq 0$ from state $p$ to state $q$ is given by $(\mathsf{M}(\mathcal{A})^n)_{p,q}$.

PROOF.– proceed by induction on $n$. ∎

The number of words of length $n$ belonging to $L = L(\mathcal{A})$ (that is, the growth function of the language; see definition 1.33, Volume 1) is given by

$$g_L(n) = \sum_{q \in F} (\mathsf{M}(\mathcal{A})^n)_{q_0, q} \, .$$

We could also count the number of accepted words of length $n$ starting from some state $r$; that is, the growth function $g_r$ of the language $\{w \in B^* \mid \delta(r, w) \in F\}$. We have a similar formula:

$$g_r(n) = \sum_{q \in F} (\mathsf{M}(\mathcal{A})^n)_{r, q} \, . \qquad [1.1]$$

Let $P \in \mathbb{Z}[X]$ be the characteristic polynomial of $\mathsf{M}(\mathcal{A})$. It is a monic polynomial of degree $d = \mathrm{Card}(Q)$. From the Cayley–Hamilton theorem, we know that $P(\mathsf{M}(\mathcal{A})) = 0$. Therefore, there exists integers $c_{d-1}, \ldots, c_0$ such that

$$\mathsf{M}(\mathcal{A})^d = \sum_{i=0}^{d-1} c_i \mathsf{M}(\mathcal{A})^i \, .$$

Multiplying both sides by $\mathsf{M}(\mathcal{A})^n$ shows that $(\mathsf{M}(\mathcal{A})^j)_{j \geq 0}$ (and thus every component of this matrix) satisfies a linear recurrence equation. From [1.1], we deduce that the sequences $(g_r(n))_{n \geq 0}$ satisfy, for all $r \in Q$, this linear recurrence equation. But of course, the initial conditions depend on the state $r$. For the same reasons, $(g_L(n))_{n \geq 0}$ satisfies the same linear recurrence equation.

REMARK 1.14.– If the DFA has $d$ states, we obtain a recurrence relation of order $d$. We will see that several automata can recognize the same language. It is, therefore, meaningful to consider a DFA with a minimal number of states. We will develop a theory of the minimal automaton in section 1.5. There exists, up to isomorphism, a unique DFA with a minimal number of states that recognize a given regular language.

Without this theory about the minimal automaton, we can still remove useless states to obtain a recurrence relation of smaller order. States that cannot be reached from the initial state, or states from which no final state can be reached are useless. So removing those states, we have a DFA (with a partial transition function) having fewer states. The resulting automaton is usually said to be *trim*.

Let us recall formally the definition of a trim DFA.

DEFINITION 1.15.– Let $\mathcal{A} = (Q, q_0, B, \delta, F)$ be a DFA. A state $q$ is *accessible* if there exists a word $w$ such that $\delta(q_0, w) = q$. A state $q$ is *coaccessible* if there exists a word $w$ such that $\delta(q, w) \in F$. The subautomaton obtained by considering only states that are accessible and coaccessible is said to be *trim*. A *sink state* (also called *dead state*) is a non-accepting state $q$ such that, for all letters $a$, $\delta(q, a) = q$. So, whenever a DFA enters a sink state, there is no way to leave it and reach an accepting state.

It is probably worth recalling a standard result about linear recurrence sequences; see, for instance, [GRA 89].

THEOREM 1.16.– Let $d \geq 1$ and $r_0, \ldots, r_{d-1} \in \mathbb{R}$. Let $(U_n)_{n \geq 0}$ be a sequence satisfying, for all $n \geq 0$,

$$U_{n+d} = r_{d-1} U_{n+d-1} + \cdots + r_0 U_n.$$

If $\alpha_1, \ldots, \alpha_t \in \mathbb{C}$ are the roots of the polynomial $X^d - r_{d-1} X^{d-1} - \cdots - r_0$ with respective multiplicities $m_1, \ldots, m_t$, then there exists polynomials $P_1, \ldots, P_t$ of degree, respectively, less than $m_1, \ldots, m_t$ and depending only on the initial conditions $U_0, \ldots, U_{d-1}$ such that

$$\forall n \geq 0,\ U_n = P_1(n)\,\alpha_1^n + \cdots + P_t(n)\,\alpha_t^n\,. \qquad [1.2]$$

## 1.3. Multidimensional alphabet

In the first section of this chapter, we have thus far considered automata over alphabets such as $\{0, 1, 2\}$ or $\{a, b\}$. But, there is no objection to considering other finite sets as alphabets such as

$$\{0,1\}^2 = \left\{ \begin{pmatrix} 0 \\ 0 \end{pmatrix}, \begin{pmatrix} 0 \\ 1 \end{pmatrix}, \begin{pmatrix} 1 \\ 0 \end{pmatrix}, \begin{pmatrix} 1 \\ 1 \end{pmatrix} \right\}.$$

With this alphabet, the corresponding recognized language is a subset of $(\{0,1\}^2)^*$. We make no distinction between pairs written horizontally or vertically, but it seems more natural to write them as a column vector because that is what the machine should read at once. Of course, there is no objection to taking $n$-tuples instead of pairs and also, we can have alphabets with more than two symbols. Note that the Cartesian product of alphabets was already considered to define the direct product of words (see definition 1.70 on page 71 of Volume 1).

EXAMPLE 1.17.– Consider the DFA depicted in Figure 1.6 (the missing transitions lead to a non-final sink). This DFA accepts pairs of words $(u, v)$ over $\{0,1\}$ of the same length where $v$ contains exactly one symbol 1 and this symbol occurs in the same position as the last 1 occurring in $u$ (and $u$ contains at least one symbol 1). Since we interchangeably use horizontal or vertical representations, let us write the two components one next to the other to fit it on a line of text. For instance, $(1,1)$, $(101010, 000010)$ or $(11100, 00100)$ belong to the recognized language. If words over $\{0,1\}$ are interpreted as base-2 expansions, then this DFA recognizes[5] pairs of positive integers $(x, y)$ (written in base 2 and possibly allowing leading zeroes) such that $y$ is the largest power of 2 dividing $x$.

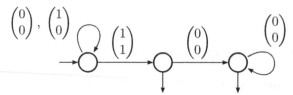

**Figure 1.6.** *A DFA accepting a language over* $\{0, 1\}^2$

EXAMPLE 1.18.– The DFA depicted in Figure 1.7 recognizes those pairs $(x, y)$ of positive integers (written in base 2 and possibly allowing leading zeroes) such that $x < y$. Again, we have not represented the sink states and the corresponding transitions.

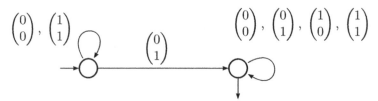

**Figure 1.7.** *A DFA accepting a language over* $\{0, 1\}^2$

EXAMPLE 1.19.– As a third example, we consider the alphabet $\{0, 1\}^3$. The DFA depicted in Figure 1.8 is intended to mimic base-2 addition. It recognizes 3-tuples of words $(u, v, w)$ such that $u$, $v$ and $w$ have the same length and $\mathrm{val}_2(u) + \mathrm{val}_2(v) = \mathrm{val}_2(w)$. The idea is that we read least significant digits first (this is not a problem because regular languages are closed under reversal). State 1 corresponds to the situation where there is a carrying to be taken into

---

5 We can say that the sequence $(V_2(n))_{n \geq 0}$ is 2-synchronized. This terminology was introduced in section 2.8, Volume 1.

account. For instance, $\mathrm{val}_2(01110) = 14$, $\mathrm{val}_2(00111) = 7$ and $\mathrm{val}_2(10101) = 21$. Reading the corresponding 3-tuples starting with the least significant digit gives the sequence

$$\begin{pmatrix}0\\1\\1\end{pmatrix} \quad \begin{pmatrix}1\\1\\0\end{pmatrix} \quad \begin{pmatrix}1\\1\\1\end{pmatrix} \quad \begin{pmatrix}1\\0\\0\end{pmatrix} \quad \begin{pmatrix}0\\0\\1\end{pmatrix}$$
$$0 \xrightarrow{\phantom{x}} 0 \xrightarrow{\phantom{x}} 1 \xrightarrow{\phantom{x}} 1 \xrightarrow{\phantom{x}} 1 \xrightarrow{\phantom{x}} 0 \, .$$

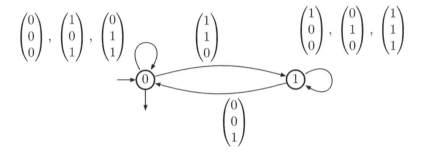

**Figure 1.8.** *A DFA accepting a language over* $\{0,1\}^3$

Let $A, B$ be finite alphabets. With these examples, we see that given an automaton over an alphabet $A \times B$, it recognizes a subset of $A^* \times B^*$, i.e. a relation over $A^* \times B^*$. In examples 1.17 and 1.19, we had moreover graphs of functions. These examples are in fact 2-synchronized functions (as briefly introduced in section 2.8, Volume 1). There is a vast amount of results about functions or relations that can be realized by finite automata or transducers, but this is not the place to take such a direction. See, for instance, [SAK 09].

## 1.4. Two pumping lemmas

The so-called pumping lemma is a classical result that is typically used to prove that some language is not regular. Its proof relies on the pigeonhole principle: any sufficiently long path goes through the same state twice.

LEMMA 1.20 (Pumping lemma).– Let $L$ be an infinite regular language. There exists an integer $k$ such that, for all words $w \in L$ of length at least $k$, there exists words $x, y, z$ where $y$ is non-empty and such that

- $w = xyz$;
- $|xy| \leq k$;
- $xy^n z \in L$ for all $n \geq 0$.

This last condition means that the language $xy^* z$ is a sublanguage of $L$.

PROOF.– Since $L$ is regular, there exists a DFA $\mathcal{A}$ with $k$ states accepting $L$. Let $w = w_1 \cdots w_\ell$ be a word of length $\ell \geq k$. Such a word exists because $L$ is infinite, and thus $L$ contains arbitrarily long words. By the pigeonhole principle, the state transition sequence of length $\ell + 1$

$$q_0, \delta(q_0, w_1), \ldots, \delta(q_0, w_1 \cdots w_i), \ldots, \delta(q_0, w_1 \cdots w_\ell)$$

contains at least twice the same state (and at least one such repetition must occur in the initial segment of length $k + 1$): there exists $i, j$ such that $0 \leq i < j \leq k$ such that

$$\delta(q_0, w_1 \cdots w_i) = \delta(q_0, w_1 \cdots w_j) .$$

It is naturally understood that if $i = 0$, then $w_1 \cdots w_0 = \varepsilon$. To complete the proof, take $x = w_1 \cdots w_i$, $y = w_{i+1} \cdots w_j$ and $z = w_{j+1} \cdots w_\ell$. We have detected a cycle labeled $y$ starting from the state $\delta(q_0, w_1 \cdots w_i)$. ∎

EXAMPLE 1.21.– The archetypical example of a non-regular language is the language $C = \{\text{a}^n \text{b}^n \mid n \geq 0\}$. Assume, to get a contradiction, that $C$ is regular. We can, therefore, make use of the pumping lemma for some constant $k$. Consider the word $w = \text{a}^k \text{b}^k$ in $C$. There must exist a factorization of $w$ of the form $xyz$ with $|xy| \leq k$. Hence, the non-empty word $y$ contains only a's. The pumping lemma implies that $xy^n z$ belongs to $C$ for any $n > 0$ leading to a contradiction (such a word has more a's than b's).

Here is another result that can be useful to prove non-regularity results. If $L$ is a regular language, then we can prove that the set of lengths

$$|L| = \{n \mid \exists w \in L : |w| = n\}$$

is a finite union of arithmetic progressions. The converse does not hold: the non-regular language $C = \{\mathtt{a}^n\mathtt{b}^n \mid n \geq 0\}$ is such that $|C| = 2\mathbb{N}$. For an example, the language $\{\mathtt{a}^{n^2} \mid n \geq 0\}$ is not a regular language.

We present and give a proof of a stronger version of the pumping lemma[6]. In contrast with lemma 1.20, we have here a necessary and sufficient condition for regularity. Note also that the following result is about any word in $B^*$ and not only words in $L$.

THEOREM 1.22 (J. Jaffe).– An infinite language $L \subseteq B^*$ is regular if and only if there exists a constant $k > 0$ such that, for all words $w \in B^*$, if $|w| \geq k$, then there exists $x, y, z \in B^*$ such that $w = xyz$ with $y$ non-empty and, for all $i \geq 0$ and all $v \in B^*$,

$$wv \in L \Leftrightarrow wy^i zv \in L. \qquad\qquad [1.3]$$

PROOF.– The fact that the condition is necessary follows the same lines as the proof of the pumping lemma. The interesting part is to prove the sufficient condition for regularity. Assume that $L$ is a language satisfying the assumptions of the statement. We consider the following DFA where

 – the set of states is $Q = \{q_w \mid w \in B^{\leq k-1}\}$;

 – the initial state is $q_\varepsilon$;

 – a state $q_w$ is final whenever $w \in L$.

---

6 There is another nice version of this result in [STA 82].

The transition function is defined as follows. If $|w| < k - 1$, then

$$\delta(q_w, b) = q_{wb}, \quad \forall b \in B.$$

So the readers may observe that this DFA is similar to a trie (as introduced in example 1.36, Volume 1) built for the (finite) prefix closed language $B^{\leq k-1}$. Let $b \in B$. If $|w| = k - 1$, then $wb$ is a word of length $k$, and we can make use of the assumption. There is at least one factorization $wb = xyz$ verifying the properties given in the statement of the result. If such a factorization is not unique, we can choose the one for which $xy$ is the shortest and then the one for which $y$ is the shortest. In that case, since $y$ is non-empty, $|xz| = k - |y| < k$, and the following definition is, therefore, legitimate:

$$\delta(q_w, b) = q_{xz}.$$

To complete the proof, we show that this DFA accepts exactly the language $L$. Proceed by induction on the length of the input word. If the input word $w$ has length less than $k$, then by definition of the DFA, this word is trivially accepted: $\delta(q_\varepsilon, w) = q_w \in F$. Let $n \geq k$. Now assume that the DFA accepts exactly the words in $L$ of length less than $n$. Let us prove that it also accepts the words in $L$ of length $n$. Let $w$ be a word of length $n$ having $p$ as prefix of length $k$: $w = pv$ for some suffix $v$. By definition of the DFA, there exists some $x, y, z \in B^*$ such that

$$\delta(q_\varepsilon, p) = q_{xz} \text{ and } p = xyz \text{ with } y \neq \varepsilon.$$

In particular, applying [1.3] for $i = 0$ means that $w$ belongs to $L$ if and only if $xzv$ belongs to $L$. In the DFA, we have

$$\delta(q_\varepsilon, w) = \delta(q_\varepsilon, pv) = \delta(q_{xz}, v).$$

This means that $w$ is accepted by the DFA if and only if $xzv$ is accepted. But observe that $|xzv| < n$ (because $y$ is non-empty). By the induction hypothesis, $xvz$ is accepted by the DFA if and only if it belongs to $L$. We conclude that $w$ is in $L$ if and only if it is accepted by the DFA. ∎

## 1.5. The minimal automaton

Infinitely many DFAs accept a given infinite regular language. Among all these automata, we seek an automaton having a minimal number of states. As we will see, up to isomorphism, this automaton is unique. Moreover, there exists important relations between a DFA and the minimal automaton accepting the same language. These relations are expressed by morphisms of automata.

The notion of minimal automaton can be defined for any language (even for a non-regular language but in that case, the automaton that we will define will be infinite).

Let $L \subseteq B^*$ be a language. Let $w$ be a word. We let $w^{-1}L$ denote the set

$$\{u \in B^* \mid wu \in L\}.$$

Such a set is called a *derivative*[7] or a *quotient*. We can, therefore, introduce an equivalence relation over $B^*$. Two words $u, v \in B^*$ are equivalent with respect to $\sim_L$ if and only if $u^{-1}L = v^{-1}L$. It is easy to see that this relation is indeed a right congruence (with respect to concatenation of words): for all $b \in B$,

$$u \sim_L v \Rightarrow ub \sim_L vb.$$

The relation $\sim_L$ is often referred to as the *Nerode congruence*.

Observe that, for all words $u, v$, we have

$$(uv)^{-1}L = v^{-1}(u^{-1}L).$$ [1.4]

---

7 There is no possible confusion with the derived sequence of a uniformly recurrent word introduced in section 3.2, Volume 1. Here, a derivative is a language of finite words.

EXAMPLE 1.23.– Let $L$ be the language made up of words $w$ over $\{a, b\}$ such $|w|_a \equiv 0 \pmod 3$. For instance, we have abbaba $\sim_L$ aaa, b $\not\sim_L$ ab, aba $\not\sim_L$ bab and a $\sim_L$ ababaa. Indeed, it is easy to see that the two words $u, v$ are such that $u \sim_L v$ if and only if $|u|_a \equiv |v|_a \pmod 3$.

Let $\mathcal{A} = (Q, q_0, B, \delta, F)$ be a deterministic[8] automaton. Let $q \in Q$ and $R$ be a subset of states. Similarly to the derivative of a language, we define $q^{-1}R$ as the set

$$\{w \in B^* \mid \delta(q, w) \in R\}.$$

In particular, the language accepted by $\mathcal{A}$ is $L(\mathcal{A}) = q_0^{-1}F$.

Let $q$ be a state and $w$ be a word such that $\delta(q_0, w) = q$. We have

$$q^{-1}F = w^{-1}L. \tag{1.5}$$

DEFINITION    1.24.– A    deterministic    automaton $\mathcal{A} = (Q, q_0, B, \delta, F)$ is *reduced* if, for all states $p, q \in Q$,

$$p^{-1}F = q^{-1}F \Rightarrow p = q.$$

This means that the languages accepted from any two distinct states differ.

DEFINITION 1.25 (Minimal automaton of $L$).– Let $L \subseteq B^*$ be a language. We define the deterministic automaton $\mathcal{A}_L = (Q_L, q_{0,L}, B, \delta_L, F_L)$ where

– the set of states is $Q_L = \{w^{-1}L \mid w \in B^*\}$;

– the initial state is $q_{0,L} = \varepsilon^{-1}L = L$;

– the set of final states is $F_L = \{w^{-1}L \mid w \in L\}$;

---

8 We do not say that this automaton is finite. The theory could be carried on for automata with infinitely many states.

– the transition function $\delta_L : Q_L \times B \to Q_L$ is defined, for all $b \in B$ and $q \in Q_L$, by

$$\delta(q, b) = b^{-1}q.$$

The transition function is well-defined. Assume that two words $u$ and $v$ are such that $u \sim_L v$, i.e. $u^{-1}L = v^{-1}L$. Then these two words correspond to the same state of $\mathcal{A}_L$. Notice that, from [1.4], we get

$$(ub)^{-1}L = b^{-1}(u^{-1}L) = b^{-1}(v^{-1}L) = (vb)^{-1}L.$$

Otherwise stated, the fact that $\sim_L$ is a right congruence implies that the value of $\delta(q, w)$ does not depend on the word $u$ such that $q = u^{-1}L$.

The function $\delta$ can be extended to $Q_L \times B^*$: for a word $w$, we get $\delta(q, w) = w^{-1}q$.

Also, $\delta_L(q_{0,L}, w) = w^{-1}L$ and $w^{-1}L$ is a final state if and only if $w$ belongs to $L$. This means that the minimal automaton of $L$ accepts the language $L$ (this is the property that we could have expected).

From this definition, $\mathcal{A}_L$ can be seen to be accessible and reduced. (Left as an exercise. See, for instance, [SAK 09].)

The fundamental result about minimal automata is the following one.

THEOREM 1.26.– Let $L$ be a language. Let $\mathcal{A}_L = (Q_L, q_{0,L}, B, \delta_L, F_L)$ be the minimal automaton of $L$. Let $\mathcal{B} = (Q, q_0, B, \delta, F)$ be a deterministic and accessible automaton accepting the same language $L$. There exists a map $\Phi : Q \to Q_L$ such that

– $\Phi$ is onto;

– $\Phi(q_0) = q_{0,L}$;

– for all $b \in B$ and $q \in Q$, $\Phi(\delta(q,b)) = \delta_L(\Phi(q),b)$;

– $\Phi(F) = F_L$.

We say that such a map $\Phi$ is a *morphism of automata*: transitions occurring in $\mathcal{A}_L$ are compatible with those occurring in $\mathcal{B}$. The fact that $\Phi$ is onto implies that if $\mathcal{B}$ is a finite automaton (hence $L$ is a regular language), then the minimal automaton of $L$ has a number of states that is less or equal to the number of states of $\mathcal{B}$. In particular, $L$ is a regular language if and only if its minimal automaton is finite. This is also equivalent (from the definition of $\mathcal{A}_L$) to say that the Nerode congruence $\sim_L$ is of finite index: the quotient $A^*/\sim_L$ is finite.

EXAMPLE 1.27.– In the upper part of Figure 1.9, we have depicted an accessible DFA accepting a*b*. In the lower part of the figure is represented the minimal automaton of the same language. The map $\Phi$ is such that $\Phi(i) = \Phi(i+3) = i+6$ for $i = 0,1,2$.

Note that if a map $\Phi : Q \to Q_L$ satisfies the properties given in the previous theorem, then $\Phi(q) = q^{-1}F$ for all $q \in Q$. Indeed, the following holds. Let $q \in Q$. Since $\mathcal{B}$ is accessible, there exists $w \in B^*$ such that $\delta(q_0,w) = q$. Then

$$\Phi(q) = \Phi(\delta(q_0,w)) = \delta_L(\Phi(q_0),w) = \delta_L(q_{0,L},w)$$

$$= w^{-1}q_{0,L} = w^{-1}L = q^{-1}F .$$

For the last equality, we have used [1.5].

COROLLARY 1.28.– Let $L$ be a language. Let $\mathcal{A}_L = (Q_L, q_{0,L}, B, \delta_L, F_L)$ be the minimal automaton of $L$. Let $\mathcal{B} = (Q, q_0, B, \delta, F)$ be a deterministic and accessible automaton accepting the same language $L$. The DFA $\mathcal{B}$ is reduced if and only if the map $\Phi : Q \to Q_L$ from theorem 1.26 is one-to-one.

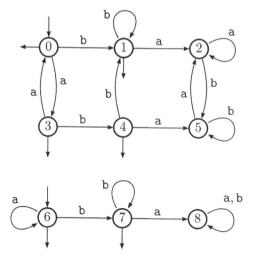

**Figure 1.9.** *An accessible DFA and the corresponding minimal automaton*

PROOF.– The map $\Phi$ given in theorem 1.26 is one-to-one if and only if, for all $p, q \in Q$, $\Phi(p) = \Phi(q)$ implies $p = q$. But, we have observed that $\Phi(p) = p^{-1}F$ and $\Phi(q) = q^{-1}F$. So, we are exactly back to the definition of a reduced automaton.   ■

The previous corollary means that, up to isomorphism, i.e. up to a bijective morphism of automata, the minimal automata of a language is unique.

COROLLARY 1.29.– Let $L$ be a regular language. Let $\mathcal{A}_L = (Q_L, q_{0,L}, B, \delta_L, F_L)$ be the minimal automaton of $L$. Let $\mathcal{B} = (Q, q_0, B, \delta, F)$ be a accessible DFA accepting the same language $L$. The minimal polynomial of the adjacency matrix of $\mathcal{A}_L$ divides the one for $\mathcal{B}$.

PROOF.– Let $A$ (respectively, $B$) be the adjacency matrix of $\mathcal{A}_L$ (respectively, $\mathcal{B}$). Consider the map $\Phi : Q \to Q_L$ given by theorem 1.26. Let $p, q$ be states in $\mathcal{A}_L$ and $w$ be a word such that $\delta_L(p, w) = q$. The key observation[9] is the following. There exists $p' \in Q$ such that $\Phi(p') = p$ and

$$(A^n)_{p,q} = \sum_{q' \in \Phi^{-1}(q)} (B^n)_{p',q'} .$$

Therefore, if $P(B) = 0$ for some polynomial $P$, then $P(A) = 0$. In particular, the minimal polynomial $M_B$ of $B$ is such that $M_B(B) = 0$. Therefore, $M_B(A) = 0$. This implies that $M_B$ is a multiple of the minimal polynomial of $A$. (This is a consequence of [1.1] on page 7 of Volume 1.)    ∎

To conclude with this section, we show that the inverse image of a regular language by a morphism is again a regular language. Note that this closure property of the set of regular languages was not listed by theorems 1.8 and 1.9.

PROPOSITION 1.30.– Let $f : A^* \to B^*$ be a morphism. Let $L \subseteq B^*$ be a regular language. The language $f^{-1}(L) \subseteq A^*$ is regular.

PROOF.– To prove that $f^{-1}(L) = \{u \in A^* \mid f(u) \in L\}$ is regular, we will show that its minimal automaton has a finite number of states.

Let $w \in A^*$. Consider languages[10] of the form $w^{-1}f^{-1}(L)$. We will show that the set

$$\{w^{-1}f^{-1}(L) \mid w \in A^*\}$$

is a finite set (proving that the corresponding minimal automaton is finite). We have

$$w^{-1}f^{-1}(L) = \{u \in A^* \mid wu \in f^{-1}(L)\}$$
$$= \{u \in A^* \mid f(wu) \in L\}$$

---

9 Think twice about it, we have to count each path of length $n$ exactly once.
10 I want to stress on the fact that the language $w^{-1}f^{-1}(L)$ may be an infinite set of words. But this is not a problem, what we are looking for is the number of distinct sets of the type $w^{-1}f^{-1}(L)$ that we can get.

$$= \{u \in A^* \mid f(w)f(u) \in L\}$$
$$= \{u \in A^* \mid f(u) \in f(w)^{-1}L\}$$
$$= f^{-1}(f(w)^{-1}L).$$

But notice that $L$ is a regular language, hence $f(w)^{-1}L$ takes a finite number of values. So, $f^{-1}(f(w)^{-1}L)$ takes only a finite number of values. ∎

What is usually understood as *state complexity* is to relate operations on languages to the size of the corresponding minimal automata. For instance, what can be said about the number of states of $A_{L \sqcup M}$ in terms of the number of states of $A_L$ and $A_M$? What kinds of lower bounds or upper bounds can be achieved? Most of the time, worst-case analysis is carried out. (But of course, an average-case analysis would be meaningful for practical reasons). We can also be interested in a family of languages depending on some parameters, let us say, some sequence of regular languages $(L_n)_{n \geq 0}$, and estimate the size of $A_{L_n}$ in terms of $n$. For a short survey, see [YU 05].

## 1.6. Some operations preserving regularity

In this section, we consider a few operations that given any regular language, will extract a sublanguage that is again regular. For these operations, we will assume that the languages are genealogically ordered (see definition 1.11 on page 14 of Volume 1). About regularity preserving operations, the paper [BER 06a] is worth reading.

Let $L$ be a language over a totally ordered alphabet $(B, <)$. Since the language is genealogically ordered (induced by the ordering of $B$), we define

$$\mathsf{maxlg}(L) = \{u \in L \mid \forall v \in L, |u| = |v| \Rightarrow u \geq v\},$$

$$\mathsf{minlg}(L) = \{u \in L \mid \forall v \in L, |u| = |v| \Rightarrow u \leq v\}.$$

Otherwise stated, maxlg($L$) (respectively, minlg($L$)) contains the largest (respectively, smallest) word of each length with respect to the genealogical ordering of $L$.

THEOREM 1.31.– Let $L$ be a regular language over a totally ordered alphabet $(B, <)$. The languages maxlg($L$) and minlg($L$) are regular.

PROOF[11].– It is convenient to make a proof exploiting non-determinism. Let $\mathcal{A}$ be a DFA accepting $L$. The idea to accept a word $w$ in maxlg($L$) is first to test if $w$ belongs to $L$ making use of $\mathcal{A}$. Next, we have to guess (we will make this statement precise) other words of the same length. The guessed word $u$ (or if you prefer, a word picked non-deterministically) is tested to be in $L$ due to $\mathcal{A}$. If we cannot guess any such word $u \in L$ larger than $w$, then $w$ belongs to maxlg($L$). Testing if $u < w$ can be realized with an automaton such as the one depicted in example 1.18.

It is easier to first build an NFA accepting $L \setminus$ maxlg($L$), and then take the complement in $B^*$ and intersection with $L$. First from $\mathcal{A}$, we construct a new DFA $\mathcal{A}_1$ accepting the language

$$L_1 = \{(u, v) \mid |u| = |v| \text{ and } u \in L\}.$$

We simply replace each label $a$ of $\mathcal{A}$ with $\mathrm{Card}(B)$ labels $\binom{a}{b}$, one for each $b \in B$. If we apply this procedure to the automaton recognizing a*b* given in Figure 1.1, we get the DFA depicted in Figure 1.10.

We do the same with

$$L_2 = \{(u, v) \mid |u| = |v| \text{ and } v \in L\}.$$

As in example 1.18, there exists a DFA accepting the language

$$L_< = \{(u, v) \in (B \times B)^* \mid u < v\}.$$

---

11 A proof of this result appeared in [SHA 94].

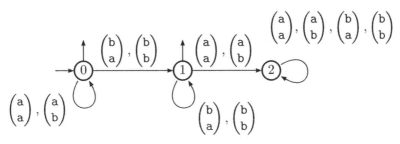

**Figure 1.10.** *A DFA accepting words in a*b* on the first component and any word of the same length on the second component*

Since regular languages are closed under intersection, the language $L_1 \cap L_2 \cap L_<$ is regular and accepted by some DFA $\mathcal{B}$ over the alphabet $B \times B$. Note that this language is exactly

$$\{(u, v) \in L \times L \mid |u| = |v| \text{ and } u < v\}.$$

If we replace every label $\binom{a}{b}$ in $\mathcal{B}$ with $a$, we get an NFA accepting $L \setminus \text{maxlg}(L)$. Indeed, if there exists an accepting path for a word $u$ in this NFA, this means that in the original DFA $\mathcal{B}$ there is at least an accepting path whose first component has label $u$. Hence, there exists a word $v$ such that $(u, v) \in L_1 \cap L_2 \cap L_<$, i.e. $u$ is in $L$ but not a maximal word in $L$. ∎

REMARK 1.32.– The last argument developed in the previous proof: replacing $\binom{a}{b}$ with $a$ is particularly important for Chapter 3, where similar constructions will be applied several times. Indeed, for the first time, we are somehow explaining how to deal with an existential quantifier.

Let $0 \leq r < m$. The *periodic decimation* of $L$ is the sublanguage obtained as follows. First, enumerate the words of $L$ in terms of increasing genealogical ordering: $L = \{w_0 < w_1 < w_2 < \cdots\}$. Then, we define

$$\text{per}_{r,m}(L) = \{w_{nm+r} \mid n \geq 0\}.$$

Indeed, this notion is similar to the periodic decimation of an infinite word as considered in proposition 2.36, Volume 1.

THEOREM 1.33.– Let $L$ be a regular language over a totally ordered alphabet $(B, <)$. Let $0 \leq r < m$. The periodic decimation $\mathrm{per}_{r,m}(L)$ is regular.

PROOF.– The first proof was originally given in [LEC 01]. Two different proofs of this result can be found with details and examples in [BER 10, pp.123–126]. See also [KRI 09] where another proof was given later.                               ∎

REMARK 1.34.– Interestingly, for a context-free language $L$ (this classical notion in formal language theory is not discussed in this book), it is known that $\mathrm{maxlg}(L)$ (respectively, $\mathrm{minlg}(L)$) is also context-free [BER 97]. Nevertheless, there exists a context-free language $C$ such that $\mathrm{per}_{0,2}(C)$ is not context-free [KRI 09].

## 1.7. Links with automatic sequences and recognizable sets

Let $b \geq 2$ be an integer. Recall that the base-$b$ expansion of $n \geq 0$ (with no leading zeroes) is denoted by $\mathrm{rep}_b(n)$. Given a set of integers $X \subseteq \mathbb{N}$, we are interested in the language $\mathrm{rep}_b(X) = \{\mathrm{rep}_b(n) \mid n \in X\}$.

DEFINITION 1.35.– Let $b \geq 2$ be an integer. A set $X \subseteq \mathbb{N}$ is $b$-recognizable if $\mathrm{rep}_b(X)$ is a regular language over $[\![0, b-1]\!]$.

EXERCISE 1.7.1.– Let $b \geq 2$. Show that $X \subseteq \mathbb{N}$ is $b$-recognizable if and only if $0^* \mathrm{rep}_b(X)$ is a regular language.

Since DFAs are a very simple model of computation, it is meaningful to study these $b$-recognizable sets. In a sense, they are "simple" from an algorithmic point of view: when written accordingly in base $b$, they are recognized by the simplest model of computation. Moreover, a DFA uses linear time in terms of the size of the input to decide whether this input belongs to the set.

REMARK 1.36.– The definition of a $b$-recognizable set can be extended to subsets of $\mathbb{N}^n$ by considering DFA over the alphabet $[\![0, b-1]\!]^n$ reading $n$-tuples of words (of the same length, the shortest ones being padded with leading zeroes) as in section 1.3. Details will be provided in definition 2.45. Such an extension will be extensively used in the last chapter of this book.

At this point of the book, the following results should be seen as easy exercises. (Hint: see proposition 2.25, Volume 1).

PROPOSITION 1.37.– Let $b \geq 2$. A set $X \subseteq \mathbb{N}$ is $b$-recognizable if and only if its characteristic sequence is $b$-automatic.

DEFINITION 1.38.– Let $\mathbf{x} \in A^{\mathbb{N}}$. The *fibers* of $\mathbf{x} = x_0 x_1 x_2 \cdots$ are sets of integers of the form

$$\mathsf{fiber}_{\mathbf{x}}(a) = \{n \in \mathbb{N} \mid x_n = a\}, \; a \in A \,.$$

PROPOSITION 1.39.– Let $a \in A$. If $\mathbf{x} \in A^{\mathbb{N}}$ is $b$-automatic, then the fiber $\mathsf{fiber}_{\mathbf{x}}(a)$ is $b$-recognizable. Conversely if, for all $a \in A$, $\mathsf{fiber}_{\mathbf{x}}(a)$ is $b$-recognizable, then $\mathbf{x} \in A^{\mathbb{N}}$ is $b$-automatic.

It is interesting to see that every divisibility criterion can be expressed in every base.

PROPOSITION 1.40.– Let $b \geq 2$. Let $m, r \in \mathbb{N}$. The arithmetic progression $m\mathbb{N} + r$ is $b$-recognizable. Hence, every ultimately periodic set of integers is also $b$-recognizable.

From proposition 1.37, we deduce that any ultimately periodic word is $k$-automatic for all $k \geq 2$.

PROOF.– Since regular languages are closed under finite modifications, we may assume that $0 \leq r < m$ and $m \geq 2$. We consider a DFA reading base-$b$ expansions most significant digit first and having $[0, m - 1]$ as set of states. Now, we define the transition function as

$$\delta(r, c) = b \cdot r + c \mod m, \quad \forall c \in [0, b - 1], r \in [0, m - 1].$$

This corresponds simply to the fact that $\mathrm{val}_b(wc) = b\, \mathrm{val}_b(w) + c$. The initial state is 0, and the final state is $r$.     ■

As an example of the construction given in the previous proof, a DFA accepting exactly the binary expansions of the integers congruent to 3 (mod 4) is given in Figure 1.11.

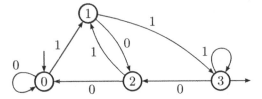

**Figure 1.11.** *A finite automaton accepting* $0^* \mathrm{rep}_2(4\mathbb{N} + 3)$

REMARK 1.41.– The above construction does not guarantee that the obtained automaton is minimal. Indeed, Alexeev studied the number of states of the minimal automaton of the language $0^* \mathrm{rep}_b(m\mathbb{N})$, that is, the set of $b$-ary representations of the multiples of $m \geq 1$ [ALE 04]. The greatest common divisor of two integers $a$ and $b$ is denoted by $\gcd(a, b)$. Let $N, M$ be such that $b^N < m \leq b^{N+1}$ and $\gcd(m, 1) < \gcd(m, b) <$ $\cdots < \gcd(m, b^M) = \gcd(m, b^{M+1}) = \gcd(m, b^{M+2}) = \cdots$. The number of states of the minimal automaton of $0^* \mathrm{rep}_b(m\mathbb{N})$ is exactly

$$\frac{m}{\gcd(m, b^{N+1})} + \sum_{t=0}^{\inf\{N, M-1\}} \frac{b^t}{\gcd(m, b^t)}.$$

We have just seen that arithmetic progressions are $b$-recognizable for all bases $b$. To see that a given infinite ordered set $X = \{x_0 < x_1 < x_2 < \cdots\}$ is $b$-recognizable for *no* base $b \geq 2$ at all, we can use results such as the contrapositive of the following theorem 1.42 where the behavior of the ratio (*respectively*, difference) of any two consecutive elements in $X$ is studied through the quantities

$$\mathbf{R}_X := \limsup_{i \to \infty} \frac{x_{i+1}}{x_i} \text{ and } \mathbf{D}_X := \limsup_{i \to \infty} (x_{i+1} - x_i).$$

Observe that if $\mathbf{D}_X$ is bounded, then this means exactly that $X$ is syndetic (we have seen this notion in definition 3.14, Volume 1).

The following "gap theorem" can be found in [COB 72] or in the classical book [EIL 74].

THEOREM 1.42.– Let $b \geq 2$. If $X$ is a $b$-recognizable infinite subset of $\mathbb{N}$, then either $\mathbf{R}_X > 1$ or $\mathbf{D}_X < +\infty$.

COROLLARY 1.43.– Let $a \in \mathbb{N}_{\geq 2}$. The set of primes and the set $\{n^a \mid n \geq 0\}$ are $b$-recognizable for no integer base $b \geq 2$.

PROOF.– Application of the above theorem for $\{n^a \mid n \geq 0\}$ is obvious. Let $\mathcal{P}$ be the set of primes. Since $n! + 2, \ldots, n! + n$ are $n-1$ consecutive composite numbers, then $\mathbf{D}_{\mathcal{P}}$ is infinite. It is known that the $n$th prime number belongs to the interval $(n \ln n, n \ln n + n \ln \ln n)$, and therefore $\mathbf{R}_{\mathcal{P}} = 1$. See, for instance, [BAC 96]. ∎

From this result, the special case of the characteristic sequence of the squares given in example 2.5, Volume 1, is $k$-automatic for no $k \geq 2$.

COROLLARY 1.44.– The characteristic sequence of the set of squares is $k$-automatic for no integer $k \geq 2$.

We conclude this section with another important theorem given to us by A. Cobham. Recall from definition 2.74, Volume 1, that two integers $p, q \geq 2$ are *multiplicatively independent* if the only integers $k, \ell$ such that $p^k = q^\ell$ are $k = \ell = 0$.

Otherwise, $p$ and $q$ are *multiplicatively dependent*. A key argument in the proof of the next statement is given in exercise 1.3.2, Volume 1.

THEOREM 1.45 (Cobham [COB 69]).– Let $p, q \geq 2$ be two multiplicatively independent integers. If a set $X \subseteq \mathbb{N}$ is both $p$-recognizable and $q$-recognizable, then $X$ is ultimately periodic.

Obviously the set $P_2 = \{2^n \mid n \geq 0\}$ of powers of two is 2-recognizable because $\mathrm{rep}_2(P_2) = 10^*$. But since $P_2$ is not ultimately periodic, the above theorem of Cobham implies that $P_2$ cannot be 3-recognizable.

EXERCISE 1.7.2.– Let $b, n \geq 2$. Show that a set $X \subseteq \mathbb{N}$ is $b$-recognizable if and only if it is $b^n$-recognizable.

As another example, we know that the Thue–Morse word t is not ultimately periodic (indeed, it is cube-free, see section 3.5, Volume 1). The set $\mathrm{fiber}_t(1)$ is the set of integers whose base-2 expansion contains an odd number of ones. This set is syndetic (because t is a concatenation of factors 01 and 10). As a result of Cobham's theorem, $\mathrm{fiber}_t(1)$ is neither 3-recognizable nor 5-recognizable.

The base dependence of recognizability shown by Cobham's result strongly motivates the general study of recognizable sets and the introduction of non-standard or exotic numeration systems discussed in Chapter 2. Defining new systems leads to new sets of integers recognizable by automata. In the last chapter of this book, we will show in particular that $b$-recognizable sets can be defined using the formalism of first-order logic.

From proposition 1.37, Cobham's theorem can be restated (in a totally equivalent way) as follows. Let $p, q \geq 2$ be two multiplicatively independent integers. If an infinite word is both $p$-automatic and $q$-automatic, then it is ultimately periodic. Therefore, instead of considering only constant-length morphisms, we could try to extend this result to morphic words. First, we have to decide what will replace the integers $p$ and $q$. This is exactly the notion of (pure) $\alpha$-substitutive word given in definition 2.71, Volume 1. Therefore, the nice extension is the Cobham–Durand theorem stated in theorem 2.75, Volume 1. To express this latter result in terms of numeration systems instead of morphic words, we will introduce abstract numeration systems.

To conclude this section, we mention an interesting extension of the notion of recognizability to sets of rational numbers [ROW 12b].

DEFINITION 1.46.– Let $b \geq 2$ be an integer. A subset $X$ of $\mathbb{Q}$ is *b-recognizable* if there exists a DFA $\mathcal{A}$ over $[\![0, b-1]\!]^2$ such that for each $x \in X$, there exists $p, q \in \mathbb{N}$ such that $x = p/q$, and for $n = \max\{|\operatorname{rep}_b(p)|, |\operatorname{rep}_b(q)|\}$, the word

$$(0^{n-|\operatorname{rep}_b(p)|} \operatorname{rep}_b(p), 0^{n-|\operatorname{rep}_b(q)|} \operatorname{rep}_b(q))$$

is accepted by $\mathcal{A}$. Moreover, any pair $(u, v)$ accepted by $\mathcal{A}$ is such that the rational number $\operatorname{val}_b(u)/\operatorname{val}_b(v)$ belongs to $X$.

The main difficulty with this definition is that a rational number is represented by infinitely many pairs of integers. But, it is only assumed that the DFA recognizes at least one of these representations.

## 1.8. Polynomial regular languages

We already know from section 1.2, that the growth function $g_L(n)$ of a regular language satisfies a linear recurrence equation (that can be derived from the characteristic polynomial of the adjacency matrix).

From the general form of the exponential polynomial given in [1.2] theorem 1.16, $g_L(n)$ exhibits either exponential or polynomial growth. This sentence should be understood as follows. If $L$ is a regular language over $B$, then the language $(B^\ell)^*$ of words having a length that is a multiple of $\ell$ is regular, and thus $L \cap (B^\ell)^*$ is again regular. As an example, assume that $g_L(n) = 2^n$ for all $n \geq 0$. Then, the growth function of $L \cap (B^\ell)^*$ is

$$g_{L \cap (B^\ell)^*}(n) = \begin{cases} 2^n, & \text{if } n \equiv 0 \pmod{\ell}; \\ 0, & \text{otherwise.} \end{cases}$$

We still say that this function has an *exponential* behavior (even if it takes the value 0 infinitely often). As a trivial example, consider the language $L = \{\text{aa}, \text{ab}, \text{ba}, \text{bb}\}^*$. Since such a phenomenon cannot be avoided, we will use the asymptotic notation $g \in \Omega_\infty(f)$ if there exists a constant $C > 0$ such that, for an infinite increasing sequence $(n_i)_{i \geq 0}$, we have

$$g(n_i) \geq Cf(n_i), \quad \forall i \geq 0.$$

So, we stress the fact that this definition is a bit different from the usual one: $g \in \Omega(f)$ if $g(n) \geq Cf(n)$ for *all* large enough $n$. Accordingly, we use $g \in \Theta'(f)$ to say that $g \in \mathcal{O}(f)$ and $g \in \Omega_\infty(f)$.

REMARK 1.47.– With regular languages, we can easily merge[12] several behaviors. Let $\ell \geq 2$. Let $L_0, \ldots, L_{\ell-1}$ be regular languages over $B$. Consider the regular language

$$L = \bigcup_{i=0}^{\ell-1} \left( L_i \cap ((B^\ell)^* B^i) \right).$$

Then, we have $g_L(n\ell + i) = g_{L_i}(n\ell + i)$ for all $i \in [\![0, \ell-1]\!]$ and $n \geq 0$.

---

12 Interestingly, for formal series in $\mathbb{S}\langle\!\langle X \rangle\!\rangle$ we can also define the relevant operation of *merge* of series as follows. Let $S_0, \ldots, S_{\ell-1} \in \mathbb{S}\langle\!\langle X \rangle\!\rangle$. The merge of these series is the series $S$ defined by $(S, X^{m\ell+i}) = (S_i, X^m)$ for $i \in [\![0, \ell-1]\!]$, $m \in \mathbb{N}$. See [BER 11].

The class of regular languages splits into two subclasses according to whether the growth function is bounded by a polynomial or is an exponential function of order $2^{\Omega_\infty(n)}$. Of course, the Jordan normal form of the adjacency matrix can provide the information we are looking for. Nevertheless, we have chosen a combinatorial approach about paths in automata to present a characterization of the regular languages with polynomial growth. Namely, the organization of the cycles in the DFA permits us to determine the asymptotic growth.

The gap between polynomial and exponential languages is rendered by the following theorem. For a proof, see [SZI 92, Theorem 6].

THEOREM 1.48.– Every regular language has a growth function that is either $\mathcal{O}(n^k)$, for some integer $k$, or $2^{\Omega_\infty(n)}$.

We will focus on the languages of polynomial growth, the internal structure of the automata recognizing such languages and the corresponding regular expressions (see theorem 1.56). Their characterization follows from a series of five lemmas stated and proved below. We conclude this section with an application to morphic words.

Indeed, as we will see, the growth function depends on the structure of the DFA accepting the language. The following results are derived from and follow the presentation in [SZI 92]. Note that other approaches can be followed; see, for instance, [JUN 08]. The paper [GAW 10] is worth reading: the authors provided a polynomial time algorithm to determine whether an NFA accepts a language of polynomial or exponential growth. Given an NFA accepting a language of polynomial growth, the order of polynomial growth can also be efficiently determined. In the same spirit, a complete discussion and an algorithm for computing the growth rate of a regular language are given in [SHU 08].

### 1.8.1. *Tiered words*

Let $\mathcal{A} = (Q, q_0, B, \delta, F)$ be a DFA. For any word $w = w_0 \cdots w_{n-1}$ of length $n$ over $B$, the *state transition sequence* of $\mathcal{A}$ on $w$, or what is sometimes called the behavior, is a word over $Q$ (we add commas for readability) denoted by

$$\mathsf{STS}_{\mathcal{A}}(w) = q_0, \delta(q_0, w_0), \ldots, \delta(q_0, w_0 \cdots w_i), \ldots,$$
$$\delta(q_0, w_0 \cdots w_{n-1}) \,.$$

The philosophy of the next definition is to detect parts of the state transition sequence where the same state is repeated, and thus cycles that are taken several times consecutively. Moreover, there is an extra condition to express that it is not allowed to come back to a previously visited cycle. To have a good intuition about the following definition, see example 1.51.

DEFINITION 1.49.– Let $\mathcal{A} = (Q, q_0, B, \delta, F)$ be a DFA[13]. A word $w \in B^*$ is said to be *t-tiered*[14], $t \geq 0$, with respect to $\mathcal{A}$ if the state transition sequence of $w$ is given by

$$\mathsf{STS}_{\mathcal{A}}(w) = \alpha \beta_1^{d_1} \gamma_1 \cdots \beta_t^{d_t} \gamma_t \qquad [1.6]$$

where

1) $0 \leq |\alpha| \leq \mathrm{Card}(Q)$;

and, for all $i \in [\![1, t]\!]$,

2) $\beta_i = q_{i,0} \cdots q_{i,k_i} \in Q^*$ and $\gamma_i = q_{i,0} r_{i,1} \cdots r_{i,\ell_i} \in Q^*$, $0 \leq k_i, \ell_i < \mathrm{Card}(Q)$;

---

13 If the automaton is non-deterministic, then the state transition sequence is not uniquely defined.

14 We stick to the terminology found in [SZI 92]. From the dictionary, *tiered* (adjective): having or arranged in tiers, rows or layers; *tier* (noun): a particular level in a group, organization, etc.

3) $q_{i,0}$ appears only once, as the first state, in $\beta_i$ and $\gamma_i$, i.e.

$$q_{i,1}, \ldots, q_{i,k_i}, r_{i,1}, \ldots, r_{i,\ell_i} \neq q_{0,i}$$

and if $i \neq j$, then $q_{i,0}$ does not appear in $\beta_j$, $\gamma_j$ nor $\alpha$;

4) $d_i > 0$.

Note that $|\beta_i| \geq 1$ because $\beta_i$ starts at least with $q_{i,0}$ and if $\ell_i = 0$, then $r_{i,1} \cdots r_{i,\ell_i} = \varepsilon$ but still $|\gamma_i| = 1$.

REMARK 1.50.– Let us make a few comments. First, let us emphasize the third condition given in the previous definition. The factor $\beta_i^{d_i} \gamma_i$ in the state transition sequence reflects that a cycle of length $|\beta_i|$ is taken $d_i$ times in the automaton. But the fact that $q_{i,0}$ does not appear in $\beta_j$, $\gamma_j$ nor $\alpha$ for $j \neq i$ implies that the cycle defined by $\beta_i$ does not appear anywhere else in the state transition sequence. The state $q_{i,0}$ has never been seen before and will never be seen after in the state transition sequence. Note that if $q_{i,0}$ appeared in some $\beta_j$ or $\gamma_j$ occurring later on, $j > i$, then we would have two cycles starting in $q_{i,0}$.

Obviously, two different words can have the same state transition sequence. Indeed, it could be the case whenever there exists two distinct letters $a, b$ such that $\delta(q, a) = \delta(q, b)$ for some state $q$.

Nevertheless, observe that if $\mathsf{STS}_{\mathcal{A}}(u) \neq \mathsf{STS}_{\mathcal{A}}(v)$, since the automaton $\mathcal{A}$ is deterministic, then $u \neq v$. We will often make use of this observation.

Observe also that $|\mathsf{STS}_{\mathcal{A}}(u)| = |u| + 1$.

EXAMPLE 1.51.– Consider the trim automaton (see definition 1.15) depicted in Figure 1.12. We have

$$\mathsf{STS}_{\mathcal{A}}(\text{bbbbbabba}) = 0\,\underline{1}313\,\underline{1}\,222\underline{2}\,2 = \alpha\beta_1^2\gamma_1\beta_2^3\gamma_2$$

with $\alpha = 0$, $\beta_1 = 13$, $\gamma_1 = 1$, $\beta_2 = 2$, $\gamma_2 = 2$. This means that after reading the first symbol, we enter a cycle of length $|\beta_1| = 2$ starting in state 1. We follow this cycle twice (the exponent of $\beta_1$). Then, we follow a path of length $|\gamma_1| = 1$ to enter the second cycle of length $|\beta_2| = 1$. We follow this cycle three times. Then, we have the last state given by $\gamma_2$. Otherwise stated, the word bbbbbabba is 2-tiered. Note that 1 (respectively, 2) occurs as the first state in $\beta_1$ and $\gamma_1$ (respectively, $\beta_2$ and $\gamma_2$) and does not appear anywhere else. Note that the word bbc̲bbabba gives exactly the same state transition sequence.

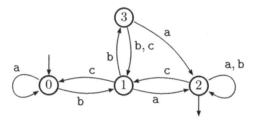

**Figure 1.12.** *A (trim) DFA*

As another example (where $\alpha = \varepsilon$), take the word aaabbacbac. We get the state transition sequence

$$0000\,132132\,1 = \beta_1^3 \gamma_1 \beta_2^2 \gamma_2$$

with $\beta_1 = 0$, $\gamma_1 = 0$, $\beta_2 = 132$, $\gamma_2 = 1$. The word bcbbb is also 2-tiered: $\text{STS}_A(\text{bcbbba}) = (01)0(13)12$. In particular, this factorization (and the detection of cycles) shows that the language $(bc)^*b(bb)^*a$ is a sublanguage accepted by $A$. Note that the growth function of this sublanguage satisfies $g(2n) = n$ and $g(2n+1) = 0$, thus $g \in \Omega_\infty(n)$. The word bbbbbacaca is also 2-tiered. But we have to be careful in the choice of the factorization:

$$\text{STS}_A(\text{bbbbbacaca}) = 01\underbrace{31}_{\beta_1}\underbrace{31}_{\gamma_1}\underbrace{2121}_{\beta_2^2}\underbrace{2}_{\gamma_2}\ .$$

This factorization shows that bb(bb)*ba(ca)* is a sublanguage accepted by the automaton. Nevertheless, we could also consider the factorization $\alpha' \beta_1'^2 \gamma_1' \beta_2'^2 \gamma_2 = 0(13)^2 1(21)^2 2$, but here the state 1 is the first state in $\beta_1'$ and $\gamma_1'$, but it also appears in $\beta_2'$ that is not allowed (third condition in definition 1.49). Actually, if we face such a factorization where the state 1 reappears after $\beta_1$ and $\gamma_1$, then this means that we have two cycles attached on state 1. If this is the case, this implies an exponential growth. In this example, looking at the state transition sequence, we see that there are two cycles with respective labels bb and ac starting from state 1. Therefore, the automaton accepts the words b(bb)$^m$(ac)$^n$a for all $m, n$. There are $2^{m+n}$ words of this form, and their length is $2(m + n + 1)$, i.e. $g_L(\ell) \geq 2^{\ell/2-1}$ for all even $\ell \geq 2$. This observation turns out to be general.

We would like to characterize regular languages with a polynomial growth, we do hope that such an example makes clear the conditions defining a tiered word. Let us summarize roughly what we have observed. As soon as we detect two cycles with distinct labels attached to state 1, then we obtain a sublanguage of exponential growth. However, if we detect a path with several consecutive cycles and we never return to a state visited at an earlier stage, then we obtain a sublanguage of polynomial growth.

### 1.8.2. *Characterization of regular languages of polynomial growth*

LEMMA 1.52.– [SZI 92, Lemma 1] Let $\mathcal{A} = (Q, q_0, B, \delta, F)$ be a DFA. If there exists a word $w \in L(\mathcal{A})$, which is $(t + 1)$-tiered, for some $t \geq 0$, with respect to $\mathcal{A}$ then the growth function of $L(\mathcal{A})$ is $\Omega_\infty(n^t)$.

PROOF.– Assume that $w \in L$ is $(t + 1)$-tiered with respect to $\mathcal{A}$. Therefore, $w$ can be factored with respect to [1.6]:

$$w = x \, y_1^{d_1} \, z_1 \cdots y_{t+1}^{d_{t+1}} \, z_{t+1}, \text{ with } y_1, \ldots, y_{t+1} \neq \varepsilon \,.$$

In particular, this means that we have detected cycles in $\mathcal{A}$: for all $j \in [\![0, t]\!]$, we have

$$\delta(q_0, x\, y_1^{d_1}\, z_1 \cdots y_{j+1}) = \delta(q_0, x\, y_1^{d_1}\, z_1 \cdots y_{j+1}^n), \ \forall n > 0 \,.$$

Let $C = |y_1| \cdots |y_{t+1}|$ and $C_i = C/|y_i|$ for $i \in [\![1, t+1]\!]$. For all $i > 0$, we set

$$n_i = |x z_1 \cdots z_{t+1}| + i\, C \,.$$

For any $t + 1$ arbitrary non-negative integers $i_1, \ldots, i_{t+1}$ such that $i_1 + \cdots + i_{t+1} = i$, the word

$$x\, y_1^{i_1 C_1}\, z_1 \cdots y_{t+1}^{i_{t+1} C_{t+1}}\, z_{t+1}$$

belongs to $L$ and it is easy to check that its length is exactly $n_i$.

Let $(i_1, \ldots, i_{t+1}) \neq (j_1, \ldots, j_{t+1})$ be two $(t+1)$-tuples of non-negative integers such that $i_1 + \cdots + i_{t+1} = i = j_1 + \cdots + j_{t+1}$. We show that the words

$$w_i = x\, y_1^{i_1 C_1}\, z_1 \cdots y_{t+1}^{i_{t+1} C_{t+1}}\, z_{t+1} \text{ and } w_j = x\, y_1^{j_1 C_1}\, z_1 \cdots y_{t+1}^{j_{t+1} C_{t+1}}$$

are distinct. There exists a smallest index $r$ such that $i_r \neq j_r$. Assume that $i_r < j_r$. Then, due to the third condition in the definition of a $(t + 1)$-tiered word, the state $q_{i_r,0} = \delta(q_0, x\, y_1^{i_1 C_1}\, z_1 \cdots y_{r-1}^{i_{r-1} C_{t+1}}\, z_{r-1})$ appears more often in $\mathsf{STS}_\mathcal{A}(w_j)$ than in $\mathsf{STS}_\mathcal{A}(w_i)$. If the state transition sequences are different, the words $w_i$ and $w_j$ are distinct, because $\mathcal{A}$ is deterministic.

Hence, the number of words of length $n_i$ belonging to $L$ is at least equal to the cardinality of the set

$$\left\{ (i_1, \ldots, i_{t+1}) \in \mathbb{N}^{t+1} : \sum_{j=1}^{t+1} i_j = i \right\},$$

and it is a classical[15] combinatorial exercise to show that the cardinality of this set is exactly

$$\binom{i+t}{t} = \frac{(i+t)\cdots(i+1)}{t!}.$$

This latter quantity is sometimes called the number of *weak compositions* of $i$. This quantity being larger than $(i+1)^t/t!$, the conclusion follows, since $n_i$ is a linear function of $i$. We have, for all large enough $i$,

$$g_L(n_i) > \frac{\left(\frac{n_i - |xz_1\ldots z_{t+1}|}{C} + 1\right)^t}{t!}.$$

∎

LEMMA 1.53.– [SZI 92, Lemma 2] Let $L$ be a regular language accepted by some DFA $\mathcal{A}$. If the growth function of $L$ is $\mathcal{O}(n^k)$ for some integer $k \geq 0$, then each word of $L$ is $t$-tiered with respect to $\mathcal{A}$ for some non-negative $t \leq k+1$.

PROOF.– The proof is by induction on the length of the word $w \in L$. If $|w| < \mathrm{Card}(Q)$, then the result holds trivially. We can write $\mathrm{STS}_{\mathcal{A}}(w)$ as $\alpha$, and $w$ is 0-tiered.

Now assume that the statement holds for all words in $L$ of length less than $n$ for some $n \geq \mathrm{Card}(Q)$. Consider a word $w$ in $L$ of length exactly $n$. Take $\mathrm{STS}_{\mathcal{A}}(w)$. By the pigeonhole principle, there exists a state $q$ such that $\mathrm{STS}_{\mathcal{A}}(w) = \mu q \nu q \tau$ where in $\nu q \tau$ all states are pairwise distinct (we look for the last repetition of a state). Consider all the occurrences of $q$ in $\mathrm{STS}_{\mathcal{A}}(w)$ and denote them by $q^{(1)}, \ldots, q^{(h+1)}$. We have $h \geq 1$. We can factor the word $w$ accordingly:

$$w = w_0 v_1 \cdots v_h w_1 \tag{1.7}$$

---

15 See for instance, theorem 29 on page 80 and exercise 87 on page 102 in [COH 78].

and $\delta(q_0, w_0) = q = \delta(q, v_i)$ for all $i \in [\![1, h]\!]$. In particular, $|w_1| = |\tau|$ and $|\nu q \tau| \leq \text{Card}(Q)$. Therefore, for all non-negative integers $n_1, \ldots, n_h$, we have

$$w_0 v_1^{n_1} \cdots v_h^{n_h} w_1 \in L.$$

Note that for any permutation $\pi$ of $[\![1, h]\!]$, the word $w_0 v_{\pi(1)}^{n_1} \cdots v_{\pi(h)}^{n_h} w_1$ also belongs to $L$ because of the repetition of the state $q$.

We prove that $v_1 = \cdots = v_h$. Proceed by contradiction and assume that there exists $i, j \in [\![1, h]\!]$, $i \neq j$, such that $v_i \neq v_j$. Now $v_i$ (respectively, $v_j$) is not a prefix of $v_j$ (respectively, $v_i$) because otherwise $\text{STS}_A(w)$ would contain more than $h + 1$ occurrences of $q$. Now consider the words in $\{v_i v_j, v_j v_i\}^*$. We claim that this set is isomorphic to $\{\mathsf{a}, \mathsf{b}\}^*$. This means that there exists a bijective morphism between the two monoids. The morphism $h : \{\mathsf{a}, \mathsf{b}\}^* \to \{v_i v_j, v_j v_i\}^*$ that maps $\mathsf{a}$ to $v_i v_j$ and $\mathsf{b}$ to $v_j v_i$ is a one-to-one correspondence. The fact that $h$ is injective derives from the fact that $v_i$ (respectively, $v_j$) is not a prefix[16] of $v_j$ (respectively, $v_i$). Now the number of words of length $n|v_i v_j|$ in $\{v_i v_j, v_j v_i\}^*$ is thus $2^n$. Therefore, there are at least $2^n$ words of length $n|v_i v_j| + |w_0 w_1|$ in $L$ contradicting the assumption that $\mathsf{g}_L \in \mathcal{O}(n^k)$. Hence, $v_1 = \cdots = v_h$.

Now comes the conclusion. The word $w_0 w_1$ belongs to $L$ and $|w_0 w_1| < n$. Using the induction hypothesis, $w_0 w_1$ is $t$-tiered for some $t \leq k + 1$: $\text{STS}_A(w_0 w_1) = \alpha \beta_1^{d_1} \gamma_1 \cdots \beta_t^{d_t} \gamma_t$, and this factorization satisfies the requirements of definition 1.49. From the choice of $q$ and [1.7], $q$ appears only once in $\gamma_t$ (if

---

16 Imagine, for instance, that $v_i = u$ and $v_j = uu$ for some word $u$. Then, clearly $h$ is not injective. For a rigorous proof, proceed by contradiction and assume that $h$ is not injective. There exists two words $x, y \in \{\mathsf{a}, \mathsf{b}\}^*$ such that $h(x) = h(y)$. We have $|x| = |y|$. There exists a smallest $k$ such that $x_k \neq y_k$. Hence, we must have $v_j v_i = v_i v_j$. We deduce that $v_i$ and $v_j$ are powers of the same word that contradicts the fact that $v_i$ (respectively, $v_j$) is not a prefix of $v_j$ (respectively, $v_i$).

$t \geq 1$, otherwise $t = 0$ and $\mathsf{STS}_{\mathcal{A}}(w_0 w_1) = \alpha$ and $q$ appears once in $\alpha$). So, we write $\gamma_t = \gamma_t' q \gamma_t''$. Hence, we get

$$\mathsf{STS}_{\mathcal{A}}(w) = \alpha \beta_1^{d_1} \gamma_1 \cdots \beta_t^{d_t} \gamma_t' \beta_{t+1}^h \gamma_t''$$

where $\beta_{t+1}$ is the state transition sequence obtained from $q$ when reading $v_1$. To see that $w$ is $t$-tiered for some $t \leq k + 1$, there are still two conditions to check. Let $i \in [\![1, t]\!]$. Let $q_{i,0}$ be the first state in $\beta_i$ and $\gamma_i$. First, observe that $q_{i,0}$ does not appear in $\beta_{t+1}$ because if it were the case, we would have two cycles starting from $q_{i,0}$. As shown at the end of example 1.51, this would imply an exponential growth not compatible with the assumption. Secondly, we must have $t < k + 1$. Indeed if $t = k + 1$, then we have found a $(k + 2)$-tiered word, and the previous lemma implies that the growth function is in $\Omega_\infty(n^{k+1})$ contradicting the assumption about the asymptotic behavior of the growth function.    ∎

LEMMA 1.54.– [SZI 92, Lemma 3] Let $\mathcal{A} = (Q, q_0, B, \delta, F)$ be a DFA. If there exists $k \geq 0$ such that each word of $L(\mathcal{A})$ is $t$-tiered with respect to $\mathcal{A}$, for some $t \leq k$, then $L(\mathcal{A})$ can be represented as a finite union of languages of the form

$$x y_1^* z_1 \cdots y_t^* z_t$$

with $0 \leq t \leq k$, $|x|, |z_1|, \ldots, |z_t| < \mathrm{Card}(Q)$ and $0 < |y_1|, \ldots, |y_t| \leq \mathrm{Card}(Q)$.

PROOF.– As observed in example 1.51, if a word $w$ in $L(\mathcal{A})$ is $t$-tiered, then it can be factored as $x y_1^{d_1} z_1 \cdots y_t^{d_t} z_t$ for some words satisfying $|x|, |z_1|, \ldots, |z_t| < \mathrm{Card}(Q)$ and $0 < |y_1|, \ldots, |y_t| \leq \mathrm{Card}(Q)$. Let $T(w) = x y_1^* z_1 \cdots y_t^* z_t$ be the corresponding language. We have $T(w) \subseteq L(\mathcal{A})$.

Therefore, we get

$$L(\mathcal{A}) = \bigcup_{w \in L} T(w).$$

At first glance, we might think that this union is infinite. But notice that the number of possible distinct languages $T(w)$ is finite. Indeed, $k$ is a constant, and $T(w)$ is formed from at most $2k + 1$ words with length bounded by $\mathrm{Card}(Q)$. ∎

LEMMA 1.55.– [SZI 92, Lemma 4] Let $t > 0$. Let $x, y_1, z_1, \ldots, y_t, z_t$ be words over $B$ where $y_1, \ldots, y_t$ are non-empty. The growth function of $L = xy_1^* z_1 \cdots y_t^* z_t$ is $\mathcal{O}(n^{t-1})$.

PROOF.– The argument is similar to that given in the proof of lemma 1.52. Let $n \geq 0$. A word $w \in L$ has length $n$ if and only if there exists $n_1, \ldots, n_t \in \mathbb{N}$ such that

$$n_1|y_1| + \cdots + n_t|y_t| + \underbrace{|xz_1 \cdots z_t|}_{C} = n \,.$$

The number of words of length $n$ in $L$ is at most the number of $t$-uples $(n_1, \ldots, n_t)$ such that $n_1|y_1| + \cdots + n_t|y_t| = n - C$. Consider the map

$$f : \mathbb{N}^t \to \mathbb{N}^t, \ (n_1, \ldots, n_t) \mapsto (n_1|y_1|, \ldots, n_t|y_t|) \,.$$

This map is clearly one-to-one. Therefore, due to the injectivity of $f$, the number of words of length $n$ in $L$ is at most

$$\mathrm{Card}\{(m_1, \ldots, m_t) \in \mathbb{N}^t \mid m_1 + \cdots + m_t = n - C\}$$

$$= \binom{n - C + t - 1}{t - 1} \,.$$

To finish the proof, observe that this quantity is in $\mathcal{O}(n^{t-1})$. ∎

In view of these lemmas, we can state the following result.

THEOREM 1.56.– A regular language $L$ is such that its growth function $g_L$ is $\mathcal{O}(n^k)$ for some $k \geq 0$ if and only if $L$ can be represented as a finite union of expressions of the form

$$xy_1^* z_1 \cdots y_t^* z_t$$

with $0 \leq t \leq k + 1$.

We have also observed that if a trim automaton accepting $L$ contains two distinct cycles (or even one cycle with different labels) attached to the same state, then the growth function is exponential. This can be reformulated as follows.

LEMMA 1.57.– [SZI 92, Lemma 6] Let $\mathcal{A} = (Q, q_0, B, \delta, F)$ be a trim DFA. Each word in $L(\mathcal{A})$ is $t$-tiered with respect to $\mathcal{A}$ for some $t \leq \text{Card}(Q)$ if and only if there does not exist any state $q \in Q$ such that there exists two distinct words $x, y$ such that $\delta(q, x) = q = \delta(q, y)$ and, for any non-trivial prefix $x'$ (respectively, $y'$) of $x$ (respectively, $y$), $\delta(q, x') \neq q$ (respectively, $\delta(q, y') \neq q$).

PROOF.– Assume that the states of $\mathcal{A}$ satisfy the above condition. Then each state in $\mathcal{A}$ appears in at most one cycle (with a unique possible label). Hence, each word in $L(\mathcal{A})$ is $t$-tiered with respect to $\mathcal{A}$ for some $t \leq \text{Card}(Q)$.

Conversely, assume that each word in $L(\mathcal{A})$ is $t$-tiered with respect to $\mathcal{A}$ for some $t \leq \text{Card}(Q)$. From lemmas 1.54 and 1.55, we know that the growth function of $L(\mathcal{A})$ is in $\mathcal{O}(n^{\text{Card}(Q)-1})$. Proceed by contradiction and assume that there exists a state in $\mathcal{A}$ from which two cycles start. Then, we already have observed that the growth function should be exponential, which is a contradiction.　∎

### 1.8.3. *Growing letters in morphic words*

We conclude this section with an application to morphic words. Let $f : A^* \to A^*$ be a morphism. With $f$ associate a matrix $\mathsf{M}_f$ and an automaton $\mathcal{A}_f$ where all states are final. See, for instance, examples 2.24 and 2.31, Volume 1. If $f$ is a constant-length morphism, then the DFA $\mathcal{A}_f$ has a total transition function. Otherwise, the transition function is partial.

What is necessary for making the connection between morphisms and regular languages is the following observation.

PROPOSITION 1.58.– For all $a \in A$, $|f^n(a)|$ is exactly the number of words of length $n$ starting from state $a$ accepted by the DFA $\mathcal{A}_f$ associated with $f$.

This can easily be shown by induction on $n$. We have to recall that the transitions of the automaton are directly obtained from the images $f(b)$, $b \in A$. Note that since every state is final, we simply have to count the number of paths of length $n$ starting from $a$.

Recall (as defined in section 3.1, Volume 1) that a letter $b$ is *growing* if $\lim_{n \to +\infty} |f^n(b)| = +\infty$. With the above discussion, we can determine which letters have polynomial growth.

EXAMPLE 1.59.– Take the morphism $f : a \mapsto ab$, $b \mapsto bc$, $c \mapsto c$. For $n \geq 1$, we can see that

$$|f^n(a)| = n(n+1)/2 + 1, \quad |f^n(b)| = n+1, \quad |f^n(c)| = 1.$$

Even though we know the exact behavior of $(|f^n(a)|)_{n \geq 0}$, $a \in \{a, b, c\}$, we will illustrate how the characterization of languages of polynomial growth can help us to get some information. We have[17] depicted the corresponding matrix $M_f$ and the associated automaton in Figure 1.13.

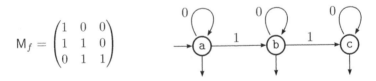

$$M_f = \begin{pmatrix} 1 & 0 & 0 \\ 1 & 1 & 0 \\ 0 & 1 & 1 \end{pmatrix}$$

**Figure 1.13.** *A DFA associated with a polynomial morphism*

17 We can also use the fact that the dominating eigenvalue of $M_f$ is 1.

The language accepted from a is $0^*10^*10^* \cup 0^*10^* \cup 0^*$. In this language, there exists a word that is 3-tiered, e.g. 01010: STS(01010) = aabbcc. From lemma 1.52, we know that $|f^n(\mathsf{a})|$ is in $\Omega_\infty(n^2)$. Moreover, from lemma 1.55, we also have that $|f^n(\mathsf{a})|$ is in $\mathcal{O}(n^2)$. The language accepted from b (respectively, c) is $0^*10^* \cup 0^*$ (respectively, $0^*$). With the same reasoning, we conclude that $|f^n(\mathsf{b})|$ (respectively, $|f^n(\mathsf{c})|$) is in $\Theta'(n)$ (respectively, $\Theta'(1)$).

EXAMPLE 1.60.– We modify the previous example a little. We just add a letter a in the image of b by the morphism. Take the morphism $f' : \mathsf{a} \mapsto \mathsf{ab}, \mathsf{b} \mapsto \mathsf{bca}, \mathsf{c} \mapsto \mathsf{ca}$. We have depicted the corresponding matrix $\mathsf{M}_{f'}$ and associated automaton in Figure 1.14.

$$\mathsf{M}_{f'} = \begin{pmatrix} 1 & 1 & 0 \\ 1 & 1 & 0 \\ 0 & 1 & 1 \end{pmatrix}$$

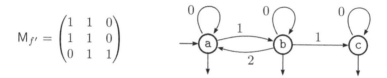

**Figure 1.14.** *A DFA associated with an exponential morphism*

We observe directly that we have two cycles originating from a (and the same holds for b). Hence, $|f'^n(\mathsf{a})|$ and $|f'^n(\mathsf{b})|$ have exponential growth. Nevertheless, words starting from c are all 1-tiered, and thus we still have $|f'^n(\mathsf{c})|$ in $\Theta'(1)$.

## 1.9. Bibliographic notes and comments

About the **history** of automata theory, see [PER 95] (in French).

We have seen that a language $L$ is regular if and only if the Nerode congruence $\sim_L$ has finite index. There is another equivalence relation that shares the same kind of property. Let $L \subseteq B^*$ be a language. The *context* of a word $u$ is the set

$C_L(u)$ of pairs $(x, y)$ of words such that $xuy$ belongs to $u$. We define the equivalence relation $\equiv_L$ by $u \equiv_L v$ if and only if $C_L(u) = C_L(v)$. This equivalence is trivially a congruence and is called **syntactic congruence**. The quotient set $B^*/ \equiv_L$ can be equipped with a product operation to get a monoid: the **syntactic monoid** of $L$. We can show that $L$ is regular if and only if the syntactic congruence $\equiv_L$ has finite index, that is, if and only if the syntactic monoid of $L$ is finite. Just like we can study state complexity (with respect to the number of states of the minimal automaton), we can also study the *syntactic complexity*: the number of equivalence classes for $\equiv_L$. For application of this concept to $b$-recognizable sets, see [LAC 12], where we compute the syntactic complexity of ultimately periodic sets of integers written in base $b$.

For **relations** and functions computed by automata; see, for instance, [FRO 93] and some chapters in [SAK 09].

There is an excellent survey about $b$-**recognizable sets** that served me since the beginning of my research work. You can consult [BRU 94]. I myself have written a survey about numeration systems giving other directions in [RIG 10].

Regarding remark 1.41 and Alexeev's result: under some mild assumptions, the state complexity of the trim minimal automaton accepting the greedy representations of the multiples of $m \geq 2$ for a wide class of linear numeration systems is studied in [CHA 11]. As an example, the number of states of the trim minimal automaton accepting the greedy representations of $m\mathbb{N}$ in the Fibonacci system is exactly $2m^2$. This will be illustrated in example 2.49.

As regards deciding whether or not a $b$-recognizable set is ultimately periodic, a fast decision procedure in $\mathcal{O}(n \log n)$ was recently presented in [MAR 13]. For subsets of $\mathbb{N}^d$, a result of Leroux was obtained earlier with quadratic complexity [LER 05].

Regarding **Cobham's theorem** for integer bases, there is a subtle mistake that is repeated in several papers (thankfully not in the original paper of Cobham). See [RIG 06] for a patch and a complete description.

A language $L$ such that $g_L(n) \in \mathcal{O}(1)$ is said to be *slender*. Regular slender languages are special cases of languages with **polynomial growth**. From the results given in section 1.8, it can be shown that a regular language is slender if and only if it is a finite union of the form $xy^*z$. See, for instance, [PĂU 95] or [SHA 94]. For periodic decimations within a slender language, see [CHA 08, Theorem 3].

We have not considered finite automata accepting **infinite words**. This subject is extensively studied. For instance, a Büchi automaton is a finite automaton devised to accept (or reject) infinite words [BÜC 60]. The acceptance condition is that a run, i.e. an infinite sequence of states corresponding to the considered infinite word, has to go through some final state infinitely often. Other acceptance conditions are also considered in the literature. To give a few references, see [PER 04]. There are several surveys [THO 90]. Some chapters of [KHO 01] are devoted to automata for infinite words.

To conclude this section, let us mention **Černý's conjecture**. Let $\mathcal{A} = (Q, q_0, B, \delta, F)$ be a DFA. Here, the initial state and the set of final states do not matter. A word $w$ is *synchronizing* if there exists a state $r$ such that, for all states $q$, $\delta(q, w) = r$. If such a word $w$ exists, then the DFA is said to be *synchronizing*. Černý's conjecture states that if a DFA with $n$ states is synchronizing, then it admits a synchronizing word of length at most $(n - 1)^2$. The DFA depicted in Figure 1.15 admits abbbabbba as a synchronizing word.

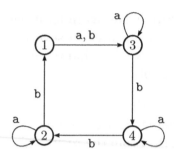

**Figure 1.15.** *A synchronizing automaton found*
*in the original paper of Černý*

There are many papers on this conjecture. A cubic bound is known. For a survey, see [VOL 08].

# A Range of Numeration Systems

> The way we do arithmetic is intimately related to the way we represent the numbers we deal with.
>
> Donald E. Knuth [KNU 12]

It is not possible (or it would deserve a complete volume) to give a comprehensive presentation of the various and exotic numeration systems that have so far been developed. Many clever non-standard systems were devised to solve relevant problems, create better or easier computational procedures, define convenient codings, etc. Numerical algorithms depend on the chosen representations. The mathematical notion of a number (e.g. integers, real numbers, complex numbers, vectors of these, ...) is independent of any chosen representation, but the way we represent or expand numbers has an influence on how operations on these numbers will be performed (e.g. comparison, addition, parallelization or pipelining of a series of computations).

EXAMPLE 2.1.– Take the usual base-2 expansions. Using the digit set $\{0, 1\}$, every positive integer $n$ has infinitely many representations of the form $0^j c_k \cdots c_0$ where $j \geq 0$, $c_k \neq 0$ and $\sum_{i=0}^{k} c_i 2^i = n$. Using the digit set $\{1, 2\}$, every positive integer has a unique expansion. Finally, with the digit set $\{-1, 0, 1\}$, we can represent positive as well as negative integers. In that case, addition of two numbers can be performed in parallel, because we can get rid of carry propagation using Avizienis's algorithm [AVI 61].

Each of these three variants of the binary system is what we call a *positional numeration system* (PNS). There exists a sequence $(U_i)_{i \geq 0}$ of integers – here the sequence $(2^i)_{i \geq 0}$ – such that every word $c_k \cdots c_0$ is a representation of the integer $\sum_{i=0}^{k} c_i U_i$. Roughly speaking, the weight of the digit $c_j$ depends on its position within the word. If the sequence $(U_i)_{i \geq 0}$ is increasing, the leftmost non-zero digit (respectively, the rightmost digit) is, thus, referred to as the *most significant* (respectively, *least significant* digit).

I have decided to give merely a short list of some systems that can be encountered. Focus is placed on systems linked to other parts of this book. In the first three sections, we will only represent integers using finite words. Several surveys about numeration systems exist; see the bibliographic notes at the end of this chapter.

To make the connection with the morphic words discussed this far, we present the Dumont–Thomas numeration system at first. We have already used this system in our discussion of the tribonacci word in section 2.2.2, Volume 1. It probably feels a bit unnatural for the reader used to the theory of numeration systems to start with such a system: the decomposition of an integer is derived from the factorization of a word and not from a sequence of integers. If the reader has skipped the first chapters about morphic words, then he/she can move directly to section 2.2.

In section 2.2, we present, in contrast to PNSs, what we call abstract numeration systems (ANS). They are primarily used to represent integers. The main idea behind these systems is that each integer has a unique representation and the set of representations of all the non-negative integers is a regular language. What matters is the ordering – or rank – of the words within the language. These systems are a generalization of integer base systems. Many "usual" numeration systems can be seen as particular ANS. To get a unified framework, we have chosen to present them almost directly. In this book, the focus is placed on infinite words. Therefore, an important part of section 2.2 is devoted to $S$-automatic words. Recall that a $k$-automatic word, as introduced in section 2.3, Volume 1, can be obtained by feeding a deterministic finite automaton with output (DFAO) with base-$k$ expansions. If we replace these expansions with representations in an abstract numeration $S$, then we get a generalization of the notion of $k$-automatic sequence. We extend Cobham's theorem on automatic words to the larger class of morphic words: an infinite word is morphic if and only if it is a "generalized" $S$-automatic word.

ANS are usually too general to obtain relevant information when dealing with arithmetic operations such as addition or multiplication by a constant. In the third section, we consider at last standard material and present (positional) numeration systems constructed from an increasing sequence of integers. This family contains integer base systems or the Fibonacci/Zeckendorf system. If the language of the numeration is a regular language, then we have special instances of ANS. In particular, this will lead us to introduce systems based on a Pisot number in section 2.4. These latter systems can be seen as very convenient PNSs: normalization and thus addition or multiplication by a constant as well as the set of representations of all the integers can be recognized by finite automata. For instance, the Zeckendorf system based on the Fibonacci sequence is a Pisot system associated with the golden ratio.

In section 2.5, we turn to the representation of real numbers and give some complementary material about the so-called $\beta$-expansions. These infinite words were introduced in the first chapter of Volume 1 to give some examples of symbolic dynamical systems. We present Parry's theorem (theorem 2.68) that permits us to describe the admissible $\beta$-expansions and the so-called $\beta$-shift. Thanks to Bertrand's theorem (theorem 2.74) a link between PNSs representing integers and $\beta$-expansions is highlighted. Recently, Ito and Sadahiro have introduced $(-\beta)$-expansions. Basics about these systems with a negative base are presented at the end of section 2.5. In particular, it is possible, using the digit-set $[\![0, \lceil \beta \rceil - 1]\!]$, to represent positive as well as negative real numbers.

In section 2.6, we give an overview of some other numeration systems that we might encounter: the rational base system, the combinatorial system, the factorial system, the Ostrowski system based on the convergents of continued fraction expansions, Cantor series expansions and shift radix systems. We give some more details about numeration systems with a rational base in $\mathbb{Q}_{>1} \setminus \mathbb{N}$ exhibiting a particular behavior. They were introduced by Akiyama, Frougny and Sakarovitch. Having a base that is neither a Pisot number nor an irrational number, i.e. a non-integer rational base, leads to some interesting phenomena.

## 2.1. Substitutive systems

We will present the Dumont–Thomas numeration system. It was already sketched in section 2.2.2, Volume 1, when dealing with some properties of the tribonacci word. It is based on morphic words (but it is not specific to the tribonacci word). The idea is to decompose, in a greedy way, a word with factors related to iterated images of letters by the morphism.

Let $f : A^* \to A^*$ be a morphism that is prolongable on a letter $a_0$. Moreover, we assume that $f$ is non-erasing. Let $\mathbf{w} = f^\omega(a_0) = w_0 w_1 w_2 \cdots$ be a fixed point of $f$.

Let $n$ be a positive integer. We will build a factorization of the prefix of length $n$ of w and this factorization will correspond to the representation of $n$ in this numeration system. Note that at the same time, we will define a finite sequence of pairs $(p_i, a^{(i)})_{i=0,\ldots,\ell}$ in $A^* \times A$ where, for all $i > 0$, $p_{i-1} a^{(i-1)}$ is a prefix of $f(a^{(i)})$. There exists a unique $\ell$ such that

$$|f^\ell(a_0)| \leq n < |f^{\ell+1}(a_0)|.$$

Otherwise stated, the word $w_0 \cdots w_{n-1}$ of length $n$ is a proper prefix of $f^{\ell+1}(a_0)$. Write

$$f(a_0) = a_{0,0} a_{0,1} \cdots a_{0,k_0} \quad \text{where } k_0 = |f(a_0)| - 1.$$

Then $f^{\ell+1}(a_0) = f^\ell(a_{0,0}) f^\ell(a_{0,1}) \cdots f^\ell(a_{0,k_0})$ and there exists $j \in [\![0, k_0 - 1]\!]$ such that

$$|f^\ell(a_{0,0} \cdots a_{0,j})| \leq n < |f^\ell(a_{0,0} \cdots a_{0,j} a_{0,j+1})|.$$

Let $p_\ell$ denote the proper prefix $a_{0,0} \cdots a_{0,j}$ of $f(a_0)$. Hence we can write

$$w_0 \cdots w_{n-1} = f^\ell(p_\ell) s$$

where $s$ is the proper prefix of $f^\ell(a^{(\ell)})$ where we set $a^{(\ell)} := a_{0,j+1}$.

We can iterate the factorization process. At step $i \in [\![1, \ell]\!]$, assume that we have a proper prefix $s$ of some $f^i(a^{(i)})$. Write $f(a^{(i)}) = a_{i,0} a_{i,1} \cdots a_{i,k_i}$ where $k_i = |f(a^{(i)})| - 1$. Two situations may occur.

1) If $|s| \geq |f^{i-1}(a_{i,0})|$, then there is a $j' \in [\![0, k_i - 1]\!]$ such that

$$s = f^{i-1}(p_{i-1}) s'$$

where $p_{i-1} := a_{i,0} \cdots a_{i,j'}$ is a proper prefix of $f(a^{(i)})$ and $s'$ is a proper prefix of $f^{i-1}(a^{(i-1)})$ where we set $a^{(i-1)} := a_{i,j'+1}$.

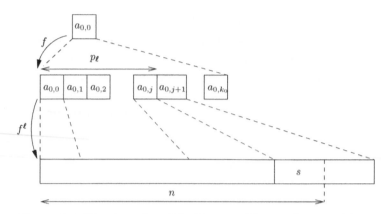

**Figure 2.1.** *First iteration of the Dumont–Thomas decomposition*

2) Otherwise, $|s| < f^{i-1}(a_{i,0})$. We set $p_i := \varepsilon$ and $a^{(i-1)} := a_{i,0}$. Thus, in particular, $s$ is a proper prefix of $f^{i-1}(a^{(i-1)})$.

At each step in this procedure, the length of the remaining word is non-increasing and the exponent $i$ is decreasing. So the procedure terminates.

Therefore, $w_0 \cdots w_{n-1}$ is decomposed as:

$$w_0 \cdots w_{n-1} = f^{\ell}(p_{\ell}) f^{\ell-1}(p_{\ell-1}) \cdots f(p_1) p_0$$

and    $n = \displaystyle\sum_{j=0}^{\ell} |f^j(p_j)|$

where, for all $i > 0$, the word $p_{i-1}$ is a proper prefix of $f(a^{(i)})$ for some $a^{(i)} \in A$. Note that if $i < \ell$, the word $p_i$ can be equal to $\varepsilon$. We will reconsider these expansions with an alternative point of view in the forthcoming example 2.20.

EXAMPLE 2.2.– Consider the Fibonacci morphism:

$$\mathcal{F} : \{0,1\}^* \to \{0,1\}^*, \ 0 \mapsto 01, \ 1 \mapsto 0 .$$

The prefix of length 12 of f is $w = 010010100100$. We have $|\mathcal{F}^4(0)| = 8 \leq 12 < |\mathcal{F}^5(0)| = 13$ and $w = \mathcal{F}^4(0)\, s$, where

$s = 0100$ is a prefix of $\mathcal{F}^4(1) = 01001$. With our notation, we have $(p_4, a^{(4)}) = (0, 1)$. Next, $\mathcal{F}(1) = 0$ and $|\mathcal{F}^3(0)| = 5 > |s|$. We have $(p_3, a^{(3)}) = (\varepsilon, 0)$. Hence, $s$ is a prefix of $\mathcal{F}^3(0) = \mathcal{F}^2(01)$ and we can write $s = \mathcal{F}^2(0)\, s' = 010\, s'$ where $s' = 0$ is a proper prefix of $\mathcal{F}^2(1) = \mathcal{F}(0) = 01$, thus $(p_2, a^{(2)}) = (0, 1)$. Finally we obtain:

$$w = \mathcal{F}^4(0)\, \mathcal{F}^3(\varepsilon)\, \mathcal{F}^2(0)\, \mathcal{F}(\varepsilon)\, 0.$$

The sequence $(p_i, a^{(i)})_{i=0,\dots,4}$ is $(0,1), (\varepsilon, 0), (0,1), (\varepsilon, 0), (0,1)$. The factorization of the prefixes of f of length 1 to 12 are given in Table 2.1.

| 1 | 0 | 7 | $\mathcal{F}^3(0)\mathcal{F}(0)$ |
|---|---|---|---|
| 2 | $\mathcal{F}(0)$ | 8 | $\mathcal{F}^4(0)$ |
| 3 | $\mathcal{F}^2(0)$ | 9 | $\mathcal{F}^4(0)\,0$ |
| 4 | $\mathcal{F}^2(0)\,0$ | 10 | $\mathcal{F}^4(0)\,\mathcal{F}(0)$ |
| 5 | $\mathcal{F}^3(0)$ | 11 | $\mathcal{F}^4(0)\,\mathcal{F}^2(0)$ |
| 6 | $\mathcal{F}^3(0)\,0$ | 12 | $\mathcal{F}^4(0)\,\mathcal{F}^2(0)\,0$ |

**Table 2.1.** *Factorization of the prefixes of* f *of length* 1 *to* 12

Note that the case of the tribonacci morphism will be treated in the generic example 2.11.

EXAMPLE 2.3.– Consider a somehow arbitrary morphism $f : a \mapsto abc, b \mapsto bc, c \mapsto ab$. The prefix of length 13 of $f^\omega(a)$ is abcbcabbcabab. This word can be factored as $f^2(ab)ab$. The word ab is a proper prefix of $f^2(c) = f(a)f(b)$. We have $(p_2, a^{(2)}) = (ab, c)$. But $|f(a)| = 3 > |ab|$, thus $(p_1, a^{(1)}) = (\varepsilon, a)$. Finally, $(p_0, a^{(0)}) = (ab, c)$. The factorization of the prefixes of $f^\omega(a)$ of length 1 to 14 are given in Table 2.2.

DEFINITION 2.4.– A finite sequence of pairs $(p_i, a^{(i)})_{i=0,\dots,\ell}$ in $A^* \times A$ is *admissible* if for all $i > 0$, $p_{i-1}a^{(i-1)}$ is a prefix of

| 1 | a | | 8 | $f^2(\mathrm{a})\mathrm{b}$ |
|---|---|---|---|---|
| 2 | ab | | 9 | $f^2(\mathrm{a})f(\mathrm{b})$ |
| 3 | $f(\mathrm{a})$ | | 10 | $f^2(\mathrm{a})f(\mathrm{b})\mathrm{a}$ |
| 4 | $f(\mathrm{a})\mathrm{b}$ | | 11 | $f^2(\mathrm{ab})$ |
| 5 | $f(\mathrm{ab})$ | | 12 | $f^2(\mathrm{ab})\mathrm{a}$ |
| 6 | $f(\mathrm{ab})\mathrm{a}$ | | 13 | $f^2(\mathrm{ab})\mathrm{ab}$ |
| 7 | $f^2(\mathrm{a})$ | | 14 | $f^2(\mathrm{ab})f(\mathrm{a})$ |

**Table 2.2.** *Factorization of the prefixes of $f^\omega(\mathrm{a})$ of length 1 to 14*

$f(a^{(i)})$. Let $a \in A$. This sequence is *a-admissible* if, moreover, $p_\ell\, a^{(\ell)}$ is a prefix of $f(a)$.

THEOREM 2.5 (Dumont–Thomas).– [DUM 89] Let $f : A^* \to A^*$ be a non-erasing morphism that is prolongable on a letter $a_0$. Let $n \geq 1$. There exist a unique $\ell$ and a unique admissible sequence $(p_i, a^{(i)})_{i=0,\ldots,\ell}$ such that $p_\ell \neq \varepsilon$, the sequence is $a_0$-admissible, and $w_0 \cdots w_{n-1} = f^\ell(p_\ell)f^{\ell-1}(p_{\ell-1}) \cdots f(p_1)p_0$.

In this context, a useful tool is the *prefix-suffix automaton* defined as follows. We are mainly interested in paths that we can follow in such a graph. This is the reason why we do not define initial or final states.

DEFINITION 2.6.– Let $f : A^* \to A^*$ be a morphism. The set of states of the prefix-suffix automaton associated with $f$ is $A$. There is a directed edge from $a$ to $b$ with label $(p, a, s)$ if and only if $f(b) = pas$.

REMARK 2.7.– Here we stick to the usual definition that we can find in the literature, but we should notice that this automaton is roughly a reversed version of the automaton $\mathcal{A}_{f,c}$ associated with a morphism $f : A^* \to A^*$ prolongable on a letter $c \in A$ (such as the automata presented in examples 2.24 and 2.31, Volume 1, and also examples 1.59 and 1.60). Indeed, if $f(b) = pas$, then there is a directed edge in the automaton $\mathcal{A}_{f,c}$ from $b$ to $a$ (so the edge is taken backward in the prefix-suffix automaton) with label $|p| - 1$. In particular, the adjacency matrix of the prefix-suffix automaton is the transpose of the one of $\mathcal{A}_{f,c}$.

Note that if $(p, a, s)$ is an edge whose destination is the vertex $b$, then $f(b) = pas$. A finite sequence $(p_i, a^{(i)}, s_i)_{i=0,\ldots,\ell}$ labels a path in the prefix-suffix automaton if and only if $f(a^{(i+1)}) = p_i a^{(i)} s_i$ for all $i \geq 0$. Note that such a condition is similar to the admissibility condition introduced in definition 2.4. Indeed, assume that we have an edge $(p_i, a^{(i)}, s_i)$ leading to the vertex $b$. Then $f(b) = p_i a^{(i)} s_i$, and every edge leaving $b$ that has a label of the form $(p_{i+1}, a^{(i+1)}, s_{i+1})$ is such that $a^{(i+1)} = b$.

EXAMPLE 2.8.– Consider the morphism $f$ given in example 2.3. The corresponding prefix-suffix automaton is depicted in Figure 2.2.

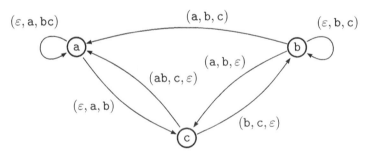

**Figure 2.2.** *The prefix-suffix automaton for*
$f : a \mapsto abc, b \mapsto bc, c \mapsto ab$

PROPOSITION 2.9.– [CAN 01] Let $f : A^* \to A^*$ be a non-erasing morphism prolongable on a letter $a_0$. For all $n$, in the prefix-suffix automaton associated with $f$, there exists a unique path $(p_i, a^{(i)}, s_i)_{i=0,\ldots,\ell}$ ending at $a_0$, i.e. $a^{(\ell)} = a_0$, such that $p_\ell \neq \varepsilon$ and $n = \sum_{j=0}^{\ell} |f^j(p_j)|$.

EXAMPLE 2.10 (Example 2.8 continued).– In the automaton depicted in Figure 2.2, consider the two paths where we have written the prefix part of each label only

$$c \xrightarrow{b} b \xrightarrow{\varepsilon} b \xrightarrow{a} a$$

and

$$b \xrightarrow{a} c \xrightarrow{ab} a .$$

They both end in a and $f$ is prolongable on a. The first (respectively, second) path gives the factorization $f^2(a)b$ (respectively, $f(ab)a$).

EXERCISE 2.1.1.– Write down a proof of proposition 2.9.

A generalization of the Fibonacci and tribonacci examples is to consider a family of morphisms as discussed in the next example. This particular example shows that integers can be decomposed using some sequence satisfying a linear recurrence equation (as will be discussed in definition 2.44).

EXAMPLE 2.11 (Generic $m$-bonacci).– Let $m \geq 1$. Let $k_0, \ldots, k_{m-1}$ be positive integers such that $k_0 \geq k_i$ for all $i > 0$.

Consider the morphism $g : [\![0, m-1]\!]^* \to [\![0, m-1]\!]^*$ such that

$$g(m-1) = 0^{k_{m-1}} \quad \text{and} \quad g(i) = 0^{k_i} (i+1) \quad \text{for } i < m .$$

In particular, for $m = 1$ and $k_0 = k_1 = 1$ (respectively, $m = 2$ and $k_0 = k_1 = k_2 = 1$), we are back to the Fibonacci word (respectively, tribonacci word). In Figure 2.3, we have represented the prefix-suffix automaton for $m = 4$. Note that to get a compact representation, we have replaced several edges with a generic label (where a parameter $j$ is involved).

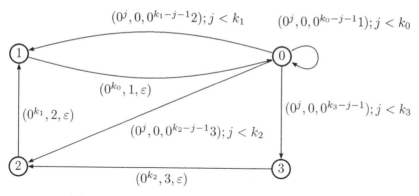

**Figure 2.3.** *The prefix-suffix automaton for* $m = 4$

The adjacency matrix associated with the prefix-suffix automaton depicted in Figure 2.3 is:

$$M = \begin{pmatrix} k_0 & k_1 & k_2 & k_3 \\ 1 & 0 & 0 & 0 \\ 0 & 1 & 0 & 0 \\ 0 & 0 & 1 & 0 \end{pmatrix}$$

which is the companion matrix of the recurrence relation:

$$U_{j+4} = k_0\, U_{j+3} + k_1\, U_{j+2} + k_2\, U_{j+1} + k_3\, U_j,\ \forall j \geq 0.$$

Indeed if $(U_j)_{j\geq0}$ satisfies this latter relation, then

$$M \begin{pmatrix} U_{j+3} \\ U_{j+2} \\ U_{j+1} \\ U_j \end{pmatrix} = \begin{pmatrix} U_{j+4} \\ U_{j+3} \\ U_{j+2} \\ U_{j+1} \end{pmatrix},\ \forall j \geq 0.$$

Instead of considering the special case $m = 4$, in a general setting, we consider the companion matrix of the recurrence relation:

$$U_{j+m} = \sum_{i=0}^{m-1} k_i\, U_{j+m-1-i},\ \forall j \geq 0. \tag{2.1}$$

We explain below that the Dumont–Thomas decomposition of every integer corresponds to a linear combination[1] of elements of such a sequence $(U_n)_{n \geq 0}$.

Let $n$ be an integer. The morphism $g$ has a very special form. Hence, if we apply the Dumont–Thomas decomposition to $g$ then we obtain

$$n = \sum_{j=0}^{\ell} |g^j(0^{c_j})| = \sum_{j=0}^{\ell} c_j |g^j(0)|, \qquad \text{[2.2]}$$

where the coefficients satisfy the following conditions:

1) $1 \leq c_\ell \leq k_0$;

2) if there exists $i$ and $r < m - 2$ such that, in the corresponding admissible sequence,

$$(p_i, a^{(i)}), (p_{i+1}, a^{(i+1)}), \ldots, (p_{i+r}, a^{(i+r)})$$
$$= (0^{k_0}, 1), (0^{k_1}, 2), \ldots, (0^{k_r}, r+1)$$

which means in particular that $c_i \cdots c_{i+r} = k_0 \cdots k_r$, then $c_{i+r+1} \leq k_{r+1}$;

3) if there exists $i$ such that, in the corresponding admissible sequence,

$$(p_i, a^{(i)}), (p_{i+1}, a^{(i+1)}), \ldots, (p_{i+r}, a^{(i+m-2)})$$
$$= (0^{k_0}, 1), (0^{k_1}, 2), \ldots, (0^{k_{m-2}}, m-1), i.e.$$

$c_i \cdots c_{i+m-2} = k_0 \cdots k_{m-2}$, then $c_{i+m-1} < k_{m-1}$.

Recall that $g(i) = 0^{k_i}(i+1)$ for $i < m - 1$. Hence, for all $r \in [\![1, m-1]\!]$, we have

$$|g^{j+r}(0)|_r = |g^j(0)|_0, \ \forall j \geq 1. \qquad \text{[2.3]}$$

---

[1] Indeed, this is a special case of a linear numeration system as introduced in definition 2.44.

Therefore, $(|g^n(0)|_0)_{n \geq 1}$ satisfies the relation [2.1] because

$$
|g^{n+m}(0)|_0
$$

$$
= k_0|g^{n+m-1}(0)|_0 + k_1|g^{n+m-1}(0)|_1 + \cdots + k_{m-1}|g^{n+m-1}(0)|_{m-1}
$$

$$
= k_0|g^{n+m-1}(0)|_0 + k_1|g^{n+m-2}(0)|_0 + \cdots + k_{m-1}|g^n(0)|_0 .
$$

The first equality comes from $|g(w)|_0 = \sum_{i=0}^{n-1} k_i |w|_i$. The second equality follows from [2.3]. For all $j \geq 1$, the first column of $M^j$ is the Parikh vector $\Psi(f^j(0))$. This can easily be proved by induction on $n$.

REMARK 2.12.– Let us make an observation about the above example (because the same ideas will be discussed when considering PNSs, see lemma 2.42). Let $d_m, \ldots, d_0$ be non-negative integers with $d_m \neq 0$. Note that if

$$
n = \sum_{j=0}^{\ell} c_j \, |g^j(0)| = \sum_{j=0}^{m} d_j \, |g^j(0)| ,
$$

where $c_\ell, \ldots, c_0$ are the coefficients obtained from the Dumont–Thomas decomposition (with $c_\ell \neq 0$) given in [2.2], then $d_m \cdots d_0 < c_\ell \cdots c_0$ where $<$ is the genealogical order given in definition 1.11. This is a result of the greediness of the Dumont–Thomas decomposition.

EXERCISE 2.1.2.– Consider a constant-length morphism and prove that the associated Dumont–Thomas system corresponds to the integer base system.

Note that the representation of real numbers is also discussed in the original paper [DUM 89].

## 2.2. Abstract numeration systems

ANS are based on an infinite regular language over a totally ordered alphabet. As we will see, they are natural generalizations of classical systems such as integer base systems or Pisot systems that will be discussed in the next section. It is easier to present some general results in this framework and then adapt them to the special cases considered in the following sections. We will mainly discuss the representation of non-negative integers (but it is also possible to represent non-negative real numbers in these systems). Recall that the genealogical ordering was introduced in definition 1.11.

DEFINITION 2.13.– An *abstract numeration system* (ANS) is a triple $S = (L, A, <)$ where $L$ is an infinite regular[2] language over a totally ordered alphabet $(A, <)$. The map $\text{rep}_S : \mathbb{N} \to L$ is the one-to-one correspondence mapping $n \in \mathbb{N}$ onto the $(n+1)$th word in the genealogically ordered language $L$, which is called the *S-representation* of $n$. The S-representation of $0$ is the first word in $L$. The inverse map is denoted by $\text{val}_S : L \to \mathbb{N}$. If $w$ is a word in $L$, $\text{val}_S(w)$ is its *S-numerical value*.

Note that $\text{val}_S(w)$ is sometimes called the *rank* of $w$. This latter notion was already introduced in section 1.4.

REMARK 2.14.– A motivation for studying ANS is that it is quite convenient to have a regular language of admissible representations. Given a finite word, we can decide in linear time with respect to the length of the entry, using a deterministic finite automaton (DFA), whether or not this word is a valid representation.

Another motivation comes from Cobham's theorem about base dependence that has been seen several times. See theorem 1.45 for its statement in terms of $k$-recognizable sets

---

2 We could relax the assumption about regularity of the language on which the numeration system is built to encompass a larger framework. Nevertheless, most of the nice properties that we will present (in particular, the equivalence with morphic words) do not hold without the regularity assumption. Recently this one-to-one correspondence was used to compress text and url addresses. See [RYO 13].

of integers and its generalization given in theorem 2.75, Volume 1, for substitutive words. In view of this theorem of Cobham, if a set of integers is recognizable within two "sufficiently different" systems, then this set is ultimately periodic. Moreover, every ultimately periodic set is always $k$-recognizable for every integer base $k \geq 2$. Therefore, if we think about a possible generalization of this theorem of Cobham, then a minimal requirement is that ultimately periodic sets – in particular $\mathbb{N}$ – should have a set of $S$-representations which is a regular language.

In contrast to PNSs, for an ANS, an integer can be decomposed using several sequences instead of a single one. We use the growth functions $g_q(n)$ for counting the number of words of length $n$ accepted from state $q$ in a DFA recognizing the language associated with the ANS [LEC 01]. The order of appearance of these functions is derived from the DFA.

THEOREM 2.15.– Let $S = (L, A, <)$ be an ANS where $L$ is accepted by the DFA $\mathcal{A} = (Q, q_0, A, \delta, F)$. Let $w = w_1 \cdots w_n \in L$. We have

$$\mathrm{val}_S(w) = \sum_{q \in Q} \sum_{i=1}^{|w|} b_{q,i}(w) \, g_q(|w| - i) \qquad [2.4]$$

where for $i = 1, \ldots, |w|$,

$$b_{q,i}(w) = \mathrm{Card}\{a \in A \mid a < w_i, \ \delta(q_0, w_1 \cdots w_{i-1}a) = q\} + \mathbf{I}_{q,q_0}$$

$$[2.5]$$

where $\mathbf{I}$ is the identity matrix in $\{0, 1\}^{Q \times Q}$, so $\mathbf{I}_{q,q_0} = 1$ if, and only if, $q = q_0$. Moreover, these coefficients are bounded:

$$0 \leq \sum_{q \in Q} b_{q,i}(w) \leq \mathrm{Card}\, A.$$

EXAMPLE 2.16 (Integer base system).– Let $b \geq 2$ be an integer. Consider the language

$$L = \{\varepsilon\} \cup \{1, \ldots, b-1\}\{0, \ldots, b-1\}^* .$$

The ANS built on $L$ using the natural ordering of the digits in $[\![0, b-1]\!]$ is the usual base-$b$ numeration system. Note that we do *not* allow leading zeroes in representations. Indeed, adding leading zeroes would change the length of the word and, therefore, the ordering (and thus the value) of this word. Recall that, in the genealogical ordering, words are first ordered with respect to their length.

EXAMPLE 2.17 (Unambiguous integer base system).– Let $b \geq 2$ be an integer. Consider the language

$$U = \{1, \ldots, b\}^* .$$

As the reader may observe, in this system, the digit set is $[\![1, b]\!]$ instead of $[\![0, b-1]\!]$. Therefore, we avoid any discussion about possible leading zeroes. For $b = 2$, the first few words in the ordered language $U$, using the natural ordering of $[\![1, b]\!]$, are

$$\varepsilon \prec 1 \prec 2 \prec 11 \prec 12 \prec 21 \prec 22 \prec 111 \prec 112 \prec 121 \prec 122 \prec \cdots .$$

Let $\mathcal{U}$ be the ANS built on $U$. Note that if $c_\ell \cdots c_0$ is a word over $[\![1, b]\!]$, then

$$\mathrm{val}_{\mathcal{U}}(c_\ell \cdots c_0) = \sum_{i=0}^{\ell} c_i \, b^i .$$

Let $n \geq 0$. In particular, note that $\mathrm{val}_{\mathcal{U}}(b^n) = b\frac{b^n - 1}{b-1}$ and the next word in the genealogical ordering, i.e. the first word of the next length, gives $\mathrm{val}_{\mathcal{U}}(1^{n+1}) = \frac{b^{n+1} - 1}{b-1}$. For more about unambiguous systems, see [HON 84, HON 92].

EXAMPLE 2.18.– Consider $L = $ a$^*$b$^*$ with a $<$ b and the ANS $\mathcal{S} = (L, \{$a, b$\}, <)$. The first few words in $L$ in ascending genealogical order are

$$\varepsilon \prec \text{a} \prec \text{b} \prec \text{aa} \prec \text{ab} \prec \text{bb} \prec \text{aaa} \prec \text{aab} \prec \text{abb} \prec \text{bbb} \prec \cdots .$$

For example, $\text{val}_\mathcal{S}(\text{abb}) = 8$ and $\text{rep}_\mathcal{S}(3) = $ aa. If we consider the bijection from $L$ to $\mathbb{N}^2$ mapping the word a$^i$b$^j$ to the pair $(i, j)$, $i, j \geq 0$, it is not difficult to see (see Figure 2.4) that the genealogical ordering of $L$ corresponds to the primitive recursive Peano enumeration of $\mathbb{N}^2$, that is

$$\text{val}_\mathcal{S}(\text{a}^i\text{b}^j) = \frac{1}{2}(i+j)(i+j+1) + j = \binom{i+j+1}{2} + \binom{j}{1}. \quad [2.6]$$

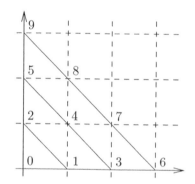

**Figure 2.4.** *Peano enumeration of* $\mathbb{N}^2$

EXAMPLE 2.19 (Prefix-closed language).– Consider the prefix-closed language

$$\{\text{a}, \text{ba}\}^*\{\varepsilon, \text{b}\}$$

of example 1.36, Volume 1. When considering such a language ordered by genealogical order, the $n$ht level of the trie (see Figure 1.4, Volume 1) contains all words of $L$ of length $n$ in lexicographic order from left to right assuming that the sons of a node are also ordered with respect to the ordering of the alphabet. To enumerate the words in the language: proceed one level at a time, from left to right.

The Dumont–Thomas numeration system discussed in the first section of this chapter turns out to be a special instance of an ANS (all states of the corresponding DFA are final).

EXAMPLE 2.20 (Dumont–Thomas numeration systems).– Every Dumont–Thomas system associated with a morphism can be interpreted as an ANS. Let $f : A^* \to A^*$ be a morphism prolongable on a letter $a \in A$. We use the same construction as in section 2.3. Consider the automaton

$$\mathcal{A}_{f,a} = (A, a, [\![0, \max_{b \in A} |f(b)| - 1]\!], \delta, A)$$

where the set of states is the alphabet $A$ and all states of $\mathcal{A}_{f,a}$ are final. The transition function $\delta : A \times [\![0, \max_{b \in A} |f(b)| - 1]\!] \to A$ is defined by $\delta(b, i) = c_i$ if $f(b) = c_0 \cdots c_{|f(b)|-1}$ and $i \in [\![0, |f(b)|-1]\!]$. If $i \geq |f(b)|$ then $\delta(b, i)$ is undefined[3]. Consider the language

$$L(\mathcal{A}) \setminus 0 [\![0, \max_{b \in A} |f(b)| - 1]\!]^*$$

and the ANS $\mathcal{S}_{f,a}$ built on this language with the genealogical ordering induced by the natural ordering of $[\![0, \max_{b \in A} |f(b)| - 1]\!]$. Note that this language is prefix-closed because every state of $\mathcal{A}_{f,a}$ is final.

Let us consider the morphism $f : a \mapsto abc, b \mapsto bc, c \mapsto ab$ prolongable on the letter $a$. This morphism was already discussed in example 2.3. In Figure 2.5, we have depicted the corresponding DFA $\mathcal{A}_{f,a}$. As explained in remark 2.7, this DFA is a reversed version of the prefix-suffix automaton given in Figure 2.2. When edges are reversed, every label $(p, a, s)$ is replaced with $|p|$.

---

3 We can add a sink/dead state to have a complete DFA. For our purposes, it is easier to work with a trim automaton.

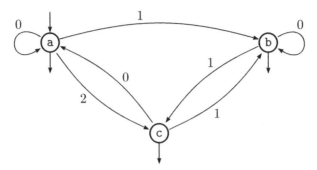

**Figure 2.5.** *The DFA* $\mathcal{A}_{f,a}$ *for* $f : \text{a} \mapsto \text{abc}, \text{b} \mapsto \text{bc}, \text{c} \mapsto \text{ab}$

The first few words in the language $L(\mathcal{A}) \setminus 0 \, [\![0, \max_{b \in A} |f(b)| - 1]\!]^*$ are listed in Figure 2.6.

| 0 | $\varepsilon$ | $\varepsilon$ |
|---|---|---|
| 1 | 1 | a |
| 2 | 2 | ab |
| 3 | 10 | $f(\text{a})$ |
| 4 | 11 | $f(\text{a})\text{b}$ |
| 5 | 20 | $f(\text{ab})$ |
| 6 | 21 | $f(\text{ab})\text{a}$ |
| 7 | 100 | $f^2(\text{a})$ |
| 8 | 101 | $f^2(\text{a})\text{b}$ |
| 9 | 110 | $f^2(\text{a})f(\text{b})$ |
| 10 | 111 | $f^2(\text{a})f(\text{b})\text{a}$ |
| 11 | 200 | $f^2(\text{ab})$ |
| 12 | 201 | $f^2(\text{ab})\text{a}$ |
| 13 | 202 | $f^2(\text{ab})\text{ab}$ |
| 14 | 210 | $f^2(\text{ab})f(\text{a})$ |
| 15 | 211 | $f^2(\text{ab})f(\text{a})\text{b}$ |
| 16 | 1000 | $f^3(\text{a})$ |

**Figure 2.6.** $\mathcal{S}_{f,a}$-*representations of the first few integers, the Dumont–Thomas expansion, and the corresponding prefix of* $f^\omega(\text{a})$

Let $c_\ell \cdots c_0$ be the $\mathcal{S}_{f,\mathsf{a}}$-representation of $n$. We can notice that the prefix of length $n$ of $f^\omega(\mathsf{a})$ is of the form

$$f^\ell(p_\ell) \cdots f(p_1) \, p_0 \text{ with } |p_i| = c_i, \ \forall i \, . \qquad [2.7]$$

Moreover, we can precisely determine the prefix $p_i$ for all $i$. It is the prefix of length $c_i$ of the word

$$f(\delta(\mathsf{a}, c_\ell \cdots c_{i+1})) \, .$$

To initialize the process, as usual, if $i = \ell$, then $c_\ell \cdots c_{i+1} = \varepsilon$ and $\delta(\mathsf{a}, \varepsilon) = \mathsf{a}$.

The observation made on this particular example is indeed general and we can compare this result with proposition 2.9.

EXERCISE 2.2.1.– Formally write the generalization of the above observation [2.7] to any Dumont–Thomas numeration system.

To conclude this section, in Table 2.3, we present an algorithm computing $\mathrm{rep}_{\mathcal{S}}(n)$ for every $n \geq 0$. Let $\mathcal{S} = (L, A, <)$ be an ANS where $L$ is accepted by the DFA $\mathcal{A} = (Q, q_0, A, \delta, F)$ and $A = \{a_1 < \cdots < a_r\}$. We use the growth functions $\mathsf{g}_q(i)$ counting the number of words of length $i$ accepted from state $q$ in $\mathcal{A}$. Note that $\sum_{i=0}^{\ell} \mathsf{g}_{q_0}(i)$ is the number of words of length at most $\ell$ in $L$.

## 2.2.1. *Generalization of Cobham's theorem on automatic sequences*

In section 2.3 Volume 1, we have considered infinite words obtained by iterating a constant-length morphism (and eventually a single application of a coding). Cobham's theorem on automatic sequences (theorem 2.28, Volume 1) states that these words obtained by iterating a morphism of length $k$ (and a coding) are exactly those that are obtained by feeding a DFAO with base-$k$ expansions. Since ANSs are a generalization of integer base numeration systems, we can also feed a DFAO with representations within an ANS. It turns out that we will generate exactly the morphic words.

```
FIND ℓ SUCH THAT ∑_{i=0}^{ℓ-1} g_{q_0}(i) ≤ n < ∑_{i=0}^{ℓ} g_{q_0}(i)
q ← q_0
m ← n - ∑_{i=0}^{ℓ-1} g_{q_0}(i)
w ← ε
FOR i=1 TO ℓ DO
    s ← 1
    WHILE m ≥ g_{δ(q,a_s)}(ℓ - i) DO
        m ← m - g_{δ(q,a_s)}(ℓ - i)
        s ← s + 1
    END-WHILE
    q ← δ(q, a_s)
    w ← wa_s
END-FOR
```

**Table 2.3.** *An algorithm for computing* $\mathrm{rep}_S(n)$

DEFINITION 2.21.– Let $S = (L, A, <)$ be an ANS. We say that an infinite word $x = x_0 x_1 x_2 \cdots \in B^{\mathbb{N}}$ is $S$-*automatic* if there exists a DFAO $(Q, q_0, A, \delta, \mu \ : \ Q \to B)$ such that $x_n = \mu(\delta(q_0, \mathrm{rep}_S(n)))$ for all $n \geq 0$.

EXAMPLE 2.22.– Let $k \geq 2$. Every $k$-automatic sequence is $S$-automatic for the ANS introduced in example 2.16.

EXAMPLE 2.23.– We consider the alphabets $A = \{a, b\}$, $B = \{0, 1, 2, 3\}$, the ANS $S = (a^* b^*, A, a < b)$ of example 2.18 and the DFAO depicted in Figure 2.7. We obtain the first few terms of the corresponding $S$-automatic sequence

$$x = 01023031200231010123023031203120231002310012 \cdots .$$

Notice that taking another ANS such as $\mathcal{R} = (\{a, ba\}^* \{\varepsilon, b\}, \{a, b\}, a < b)$, we obtain on the same DFAO, another infinite word $y = 01023\underline{1}3\underline{1}023 \cdots$ which is $\mathcal{R}$-automatic (underlined letters indicate the differences between x and y). This stresses the fact that an $\mathcal{S}$-automatic sequence really depends on two ingredients: an ANS and a DFAO.

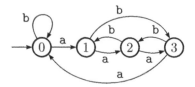

**Figure 2.7.** *A DFAO with output alphabet* $\{0, 1, 2, 3\}$

We can now state the generalization of Cobham's theorem on automatic sequences [RIG 00, RIG 02b].

THEOREM 2.24.– Let x be an infinite word. This word is morphic if and only if x is $\mathcal{S}$-automatic for some ANS $\mathcal{S}$.

This result is restated as propositions 2.29 and 2.32.

The idea of the following result is to consider the end point, whether or not it is a final state, of all paths that can be achieved in a DFA. These paths are naturally genealogically ordered with respect to their label. Recall that $\sigma$ is the shift operator introduced in definition 1.58. A special case of this result was given in [BRU 97, Prop. 18] for the Pisot systems considered in section 2.4; these systems are special cases of ANS.

LEMMA 2.25.– Let $A = \{a_1 < \cdots < a_n\}$ be a totally ordered alphabet. Let $z \notin Q$. Let $\mathcal{A} = (Q, q_0, A, \delta_\mathcal{A}, F)$ be a DFA where $\delta_\mathcal{A} : Q \times A \to Q$ is (in general) a partial function[4]. Define the

---

4 Several times we have seen the situation where $\delta$ is partial. See the first pages of sections 1.1 and 1.2. In particular, for a DFA derived from a morphism $f$, if $f$ is not a constant-length morphism, then the transition function is always partial.

morphism $\psi_{\mathcal{A}} : (Q \cup \{z\})^* \to (Q \cup \{z\})^*$ by $\psi_{\mathcal{A}}(z) = z\, q_0$ and, for all $q \in Q$,

$$\psi_{\mathcal{A}}(q) = \delta_{\mathcal{A}}(q, a_1) \cdots \delta_{\mathcal{A}}(q, a_n).$$

In the latter expression if $\delta_{\mathcal{A}}(q, a_i)$ is not defined for some $i$, then it is replaced by $\varepsilon$. Let $L$ be the regular language accepted by $(Q, q_0, \mathcal{A}, \delta, Q)$ where all states of $\mathcal{A}$ are final. Then the shifted sequence $\sigma(\psi_{\mathcal{A}}^{\omega}(z))$ is the sequence $(x_n)_{n \in \mathbb{N}}$ of the states reached in $\mathcal{A}$ by the words of $L$ in genealogical order, i.e. for all $n \in \mathbb{N}$,

$$x_n = \delta_{\mathcal{A}}(q_0, w_n)$$

where $w_n$ is the $(n+1)$th word of the genealogically ordered language $L$.

Prior to the proof, let us give a short example to set up the framework properly.

EXAMPLE 2.26.– Consider the DFA given in Figure 2.8. Note that the automaton is not complete, the transition function $\delta$ is partial: from $q_1$ we cannot read b. Assume that a $<$ b. The sequence of the ordered words in the language accepted by the automaton where all states are considered as final states are

$$(w_n)_{n \geq 0} = \varepsilon,\ \text{a, b, aa, ba, bb, aaa, aab, baa, bab, bba, aaaa}, \ldots .$$

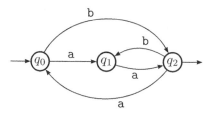

**Figure 2.8.** *A DFA*

The corresponding sequence of states is

$$(\delta(q_0, w_n))_{n \geq 0} = q_0, q_1, q_2, q_2, q_0, q_1, q_0, q_1, q_1, q_2, q_2, q_1, \ldots .$$

For instance, the second $q_2$ in the sequence, i.e. its fourth element, is the state reached by the DFA when reading aa, i.e. the fourth word $w_3$, from $q_0$. Now consider the morphism

$$\psi_\mathcal{A} : \begin{cases} z \mapsto z\, q_0 \\ q_0 \mapsto q_1\, q_2 \\ q_1 \mapsto q_2 \\ q_2 \mapsto q_0\, q_1. \end{cases}$$

We can observe that the introduction of the extra letter z gives a prolongable morphism: in this example only $\psi_\mathcal{A}(z)$ begins with z; for $x \in \{q_0, q_1, q_2\}$, the word $\psi_\mathcal{A}(x)$ does not start with $x$. Now we can compute the prefix of $\psi_\mathcal{A}^\omega(z)$,

$$\psi_\mathcal{A}^\omega(z) = z\, q_0\, q_1\, q_2\, q_2\, q_0\, q_1\, q_0\, q_1\, q_1\, q_2\, q_2\, q_1\, q_2\, q_2\, q_2 \cdots .$$

Now let us give the proof of lemma 2.25.

PROOF.–    First observe that we have the following factorization:

$$\psi_\mathcal{A}^\omega(z) = z\, x_0\, x_1\, x_2 \cdots = z\, q_0\, \psi_\mathcal{A}(q_0)\, \psi_\mathcal{A}^2(q_0) \cdots$$

and $x_0 = q_0 = \delta_\mathcal{A}(q_0, \varepsilon)$. Then by the definition of $\psi_\mathcal{A}$, if $x_n = \delta_\mathcal{A}(q_0, w_n)$, $n \geq 0$, then the factor

$$u_n = \psi_\mathcal{A}(x_n) = \delta_\mathcal{A}(q_0, w_n a_1) \cdots \delta_\mathcal{A}(q_0, w_n a_n) \qquad [2.8]$$

appears in $\psi_\mathcal{A}^\omega(z)$ with the usual convention of replacing the undefined transitions with $\varepsilon$. Indeed, $z x_0 x_1 x_2 \cdots$ is a fixed point of $\psi_\mathcal{A}$ and each $x_n$ produces a factor $\psi_\mathcal{A}(x_n) = u_n$ appearing later on in the infinite word. Moreover, this factor is preceded by

$$\delta_\mathcal{A}(q_0, w_{n-1} a_1) \cdots \delta_\mathcal{A}(q_0, w_{n-1} a_n)$$

and followed by

$$\delta_\mathcal{A}(q_0, w_{n+1} a_1) \cdots \delta_\mathcal{A}(q_0, w_{n+1} a_n).$$

Therefore, it is clear that we get all states reached from the initial state when considering the labels of all the paths in $\mathcal{A}$ in increasing genealogical order.                    ∎

EXAMPLE 2.27.– Also, the reader might feel that the introduction of the extra letter z, such that $\psi_A(\mathsf{z}) = \mathsf{z}\,q_0$, is somewhat artificial, in particular, whenever the initial state has a loop. Let us consider the following example given by the DFA depicted in Figure 2.9. Let us compare the infinite words generated by the morphism

$$\mu : \begin{cases} q_0 \mapsto q_0\,q_2 \\ q_1 \mapsto q_2 \\ q_2 \mapsto q_0\,q_1 \end{cases}$$

and by the morphism $\psi_A$ given by lemma 2.25 and defined by $\psi_A(\mathsf{z}) = \mathsf{z}q_0$ and $\psi_A(x) = \mu(x)$ for $x \in \{q_0, q_1, q_2\}$. We get

$$\psi_A^{\omega}(\mathsf{z}) = \mathsf{z}\,q_0\,q_0\,q_2\,q_0\,q_2\,q_0\,q_1\,q_0\,q_2\,q_0\,q_1\,q_0\,q_2\,q_2\,q_0\,q_2 \cdots$$

but

$$\mu^{\omega}(q_0) = q_0\,q_2\,q_0\,q_1\,q_0\,q_2\,q_2\,q_0\,q_2\,q_0\,q_1\,q_0\,q_1\,q_0\,q_2 \cdots \cdots.$$

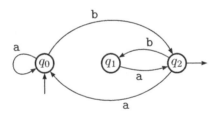

**Figure 2.9.** *Another DFA with a loop on the initial state*

We can show that the sequence $\mu^{\omega}(q_0)$ is the sequence of states reached from $q_0$ by considering only words in the DFA depicted in Figure 2.9 starting with b instead of taking into account all the possible paths. Of course, we cannot simply remove the loops of label a from $q_0$ because it may be used not solely by paths starting with a.

From the latter example, we can state the following result. In contrast to lemma 2.25, we have a stronger assumption about a loop on the initial state.

LEMMA 2.28.– Let $A = \{a_1 < \cdots < a_n\}$ be a totally ordered alphabet. Let $\mathcal{A} = (Q, q_0, A, \delta_\mathcal{A}, F)$ be a DFA where $\delta_\mathcal{A} : Q \times A \to Q$ is (in general) a partial function such that $\delta_\mathcal{A}(q_0, a_1) = q_0$ and there exists $j > 1$ such that $\delta_\mathcal{A}(q_0, a_j)$ is defined. Define the morphism $\mu_\mathcal{A} : Q^* \to Q^*$, for all $q \in Q$, by

$$\mu_\mathcal{A}(q) = \delta_\mathcal{A}(q, a_1) \cdots \delta_\mathcal{A}(q, a_n).$$

In this latter expression, if $\delta_\mathcal{A}(q, a_i)$ is not defined for some $i$, then it is replaced by $\varepsilon$. Let $M$ be the regular language accepted by $(Q, q_0, A, \delta, Q)$ where all states of $\mathcal{A}$ are final. Let $L = M \setminus a_1 A^*$. Then the sequence $\mu_\mathcal{A}^\omega(q_0)$ is the sequence $(x_n)_{n \in \mathbb{N}}$ of the states reached in $\mathcal{A}$ by the words of $L$ in genealogical order, i.e. for all $n \in \mathbb{N}$,

$$x_n = \delta_\mathcal{A}(q_0, w_n)$$

where $w_n$ is the $(n + 1)$ ht word of the genealogically ordered language $L$.

PROOF.– Left as an exercise. Adapt the proof of lemma 2.25. Note that $M$ is the set of labels of all paths starting from $q_0$ and whose first symbol is not $a_1$.    ∎

EXERCISE 2.2.2 (J. Leroy).– Let $A = \{a_1 < \cdots < a_n\}$ be a totally ordered alphabet. Let $\mathcal{A} = (Q, q_0, A, \delta_\mathcal{A}, F)$ be a DFA. Let $z$ be a symbol not in $Q$. The morphism $\mu : (Q \cup \{z\})^* \to (Q \cup \{z\})^*$ is defined, for all $q \in Q$, as in lemma 2.28 and $\mu(z) = z\,\delta_\mathcal{A}(q_0, a_1) \cdots \delta_\mathcal{A}(q_0, a_n)$. What is the sequence $\mu^\omega(z)$ ?

We know how to associate an automaton $\mathcal{A}_{f,a}$ with a morphism $f : A^* \to A^*$ prolongable on a letter $a \in A$. We have encountered this construction many times. See, for instance, examples 2.24 and 2.31, Volume 1, and also examples 1.59 and 1.60. Its formal definition was recalled in example 2.20.

If we consider a morphic word $x = g(f^\omega(a))$ where $g : A^* \to B^*$ is a coding, then we can extend $\mathcal{A}_{f,a}$ to a DFAO $\mathcal{A}_{f,a,g}$ defined as follows. We set

$$\mathcal{A}_{f,a,g} = (A, a, [\![0, \max_{b \in A} |f(b)| - 1]\!], \delta, g : A^* \to B^*).$$

A supplementary example is given after the next proposition (example 2.31).

PROPOSITION 2.29.– Let $f : A^* \to A^*$ be a morphism prolongable on the letter $a \in A$ and $g : A^* \to B^*$ be a morphism such that $x = g(f^\omega(a))$ is infinite. There exists an ANS $S$ such that $x \in B^\mathbb{N}$ is an $S$-automatic sequence. Moreover, $S$ can be explicitly derived from $f$ and $g$.

PROOF.– From Cobham's theorem on morphic words (theorem 2.12, Volume 1), we can assume that $f$ is non-erasing and that $g$ is a coding. Let $C = [\![0, \max_{b \in A} |f(b)| - 1]\!]$ and consider the DFA $\mathcal{A}_{f,a} = (A, a, C, \delta, A)$.

Let $L \subseteq C^*$ be the language recognized by $\mathcal{A}_{f,a}$. This language will be used to construct an ANS $S$ showing that $x$ is $S$-automatic. The alphabet $C$ being a subset of $\mathbb{N}$, we will consider the natural ordering of $C$. Since $f(a) = au$ for some non-empty word $u$, it is clear that if $w \in L$ then $0w \in L$. Indeed by definition of $\mathcal{A}_{f,a}$, its initial state $a$ has a loop labeled by $0$, the first letter in $C$. If we apply lemma 2.25 to this automaton $\mathcal{A}_{f,a}$, we obtain a morphism $\psi_{\mathcal{A}_{f,a}}$ generating the sequence of the states reached by the words of $L$. This morphism is defined as follows. Let $z \notin A$. We have $\psi_{\mathcal{A}_{f,a}}(z) = za$ and, for all $b \in A$, $\psi_{\mathcal{A}_{f,a}}(b) = f(b)$.

The main point leading to the conclusion is to compare $\psi^\omega_{\mathcal{A}_{f,a}}(z)$ and $f^\omega(a)$. Since $f(a) = au$, we have the following factorizations

$$f^\omega(a) = au\, f(u)\, f^2(u)\, f^3(u) \cdots$$

and

$$\psi^\omega_{\mathcal{A}_{f,a}}(\mathbf{z}) = \mathbf{z}a\,a\,u\,f(a)\,f(u)\,f^2(a)\,f^2(u)\,f^3(a)\,f^3(u)\cdots.$$

If we erase the factors $\mathbf{z}$, $a$, $f(a)$, $f^2(a)$, ... occurring in that order in the above factorization of $\psi^\omega_{\mathcal{A}_{f,a}}(\mathbf{z})$, we recover $f^\omega(a)$. Recall that $\psi^\omega_{\mathcal{A}_{f,a}}(\mathbf{z})$ is, except for $\mathbf{z}$, the sequence of states reached in $\mathcal{A}_{f,a}$ by considering all the possible paths in genealogical order. The second occurrence of $a$ in $\psi^\omega_{\mathcal{A}_{f,a}}(\mathbf{z})$ is the state reached in $\mathcal{A}_{f,a}$ when reading $0 \in L$. By the property [2.8] of $\psi_{\mathcal{A}_{f,a}}$, the factor $f^n(a)$ in the above factorization corresponds to the states reached in $\mathcal{A}_{f,a}$ when reading the words in $L$ of length $n + 1$ starting with $0$. Consequently, when giving to $\mathcal{A}_{f,a}$ the words of $L \setminus 0C^*$ in increasing genealogical order[5], we build exactly the sequence $f^\omega(a) = (y_n)_{n\geq0}$, i.e. if $w_0 \prec w_1 \prec w_2 \prec \cdots$ are the words of $L \setminus 0C^*$ in genealogical order, then $y_n = \delta(a, w_n)$ where $\delta$ is the transition function of $\mathcal{A}_{f,a}$. To finish, we have to consider the automaton $\mathcal{A}_{f,a,g}$ as a DFAO with the ANS $S$ built over $L \setminus 0C^*$ to see that the sequence $g(f^\omega(a))$ is $S$-automatic. ∎

EXERCISE    2.2.3.– Write    an    alternative    proof    of proposition 2.29 using lemma 2.28 instead of lemma 2.25.

The statement of the next result explicitly introduces the language that was built in the proof of proposition 2.29. It is a natural generalization of lemma 2.22, Volume 1. We can say that the language $L \setminus 0C^*$ is the *directive language* of $f$: if the letters in $f^\omega(a)$ are indexed by the words in $L \setminus 0C^*$, then we know precisely which letter produces which factor through the morphism.

COROLLARY 2.30.– Let $f : A^* \to A^*$ be a non-erasing morphism prolongable on the letter $a \in A$ such that $\mathbf{x} = (x_n)_{n\geq0} = f^\omega(a)$ is infinite. Consider the ANS $S$ built over

---

5 Note that this is exactly the language that we have considered in example 2.20.

$L \setminus 0C^*$ where $C = [\![0, \max_{b \in A} |\sigma(b)| - 1]\!]$ and $L$ is the language accepted by $\mathcal{A}_{f,a}$. Let $w \in L$ and $\ell \in \mathbb{N}$ be such that $|f(x_{\mathrm{val}_S(w)})| = \ell$. Then

$$f(x_{\mathrm{val}_S(w)}) = x_{\mathrm{val}_S(w0)} \cdots x_{\mathrm{val}_S(w(\ell-1))}.$$

In the above formula, for $i \in \{0, \dots, \ell - 1\}$, $wi$ has to be understood as the concatenation of $w \in L \subseteq C^*$ and $i \in C$.

PROOF[6].– This is a consequence of the proofs of lemma 2.25 and proposition 2.29. ∎

EXAMPLE 2.31.– Consider the alphabets $A = \{a, b, c\}$, $B = \{d, e\}$ and the morphisms

$$f : A^* \to A^*, \quad \begin{cases} a \mapsto abc \\ b \mapsto bc \\ c \mapsto aac. \end{cases}$$

The corresponding automaton $\mathcal{A}_{f,a,g}$ is given in Figure 2.10 and the output function is represented on the outgoing arrows. The infinite word generated by the morphism $f$ is $f^\omega(a) = (x_n)_{n \geq 0} = \text{abc}\underline{\text{b}}\text{caac}\underline{\text{bc}}\text{aacabcabcaac}\cdots$. The first few words without leading $0$ accepted by the automaton given in Figure 2.10 where all states are final are $\varepsilon$, 1, 2, 10, 11, 20, 21, 22, 100, .... This provides us with an ANS $S$.

For instance, we consider the element $x_3 = b$. This is why it has been underlined. We know that $f(b) = bc$. So the latter factor should appear later on in the infinite word and the previous corollary permits us to find where it occurs. The $S$-representation of 3 is 10. So we have to consider the words 100 and 101 – only these two words because $|f(b)| = 2$ – and $\mathrm{val}_S(100) = 8$, $\mathrm{val}_S(101) = 9$. Therefore, we can check that $x_8 x_9 = f(b) = bc$.

---

6 An independent proof is given in [BER 10, Chapter 3].

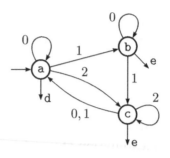

**Figure 2.10.** *The automaton* $\mathcal{A}_{f,a,g}$

Now, we turn to the converse of proposition 2.29. This is the generalization of proposition 2.26, Volume 1.

PROPOSITION 2.32.– Every $\mathcal{S}$-automatic sequence is a morphic word.

PROOF.–    Let $\mathcal{S}$ = $(L, A, <)$ be an ANS. Let $\mathcal{A}$ = $(Q, q_0, A, \delta_A, F)$ be a complete DFA accepting $L$. Let $\mathcal{B}$ = $(R, r_0, A, \delta_B, \mu : R \to B)$ be a DFAO generating an $\mathcal{S}$-automatic sequence $\mathbf{x}$ = $(x_n)_{n \geq 0}$ over $B$, i.e. for all $n \geq 0$, $x_n = \mu(\delta_B(r_0, \mathrm{rep}_\mathcal{S}(n)))$.

Consider the Cartesian product automaton $\mathcal{P}$ = $\mathcal{A} \times \mathcal{B}$ defined as follows (the construction is similar to the one given in definition 1.10). The set of states of $\mathcal{P}$ is $Q \times R$. The initial state is $(q_0, r_0)$ and the alphabet is $A$. For any word $w \in A^*$, the transition function $\Delta : (Q \times R) \times A^* \to Q \times R$ is given by

$$\Delta((q, r), w) = (\delta_A(q, w), \delta_B(r, w)) .$$

This means that the product automaton mimics in a single automaton, the behaviors of both $\mathcal{A}$ and $\mathcal{B}$. In particular, after reading $w$ in $\mathcal{P}$, $\Delta((q_0, r_0), w)$ belongs to $F \times R$ if and only if $w$ belongs to $L$. Moreover, if $\mathrm{rep}_\mathcal{S}(n) = w$ and $\Delta((q_0, r_0), w) = (q, r)$, then $x_n = \mu(r)$.

Now we can apply lemma 2.25 to $\mathcal{P}$ and define a morphism $\psi_\mathcal{P}$ prolongable on a letter z which does not belong to $Q \times R$. In view of the previous paragraph, we define $\nu : ((Q \times R) \cup \{z\})^* \to B^*$ by

$$\nu(q, r) = \begin{cases} \mu(r), & \text{if } q \in F; \\ \varepsilon, & \text{otherwise;} \end{cases}$$

and $\nu(z) = \varepsilon$. As lemma 2.25 can be used to describe the sequence of reached states, $\nu(\psi_{\mathcal{P}}(z))$ is exactly the sequence $(x_n)_{n \geq 0}$. This proves that x is morphic. ∎

Note that the morphisms obtained at the end of this proof are erasing. Again, if needed, Cobham's theorem on morphic words (theorem 2.12, Volume 1) can be used.

EXAMPLE 2.33.– Consider the ANS $\mathcal{S}$ constructed over the language a*b*. Let $\mathcal{A}$ be the minimal (complete) automaton of this language. It is depicted in the upper part of Figure 2.11. Let $\mathcal{B}$ be a DFAO. It is depicted on the left in the lower part of Figure 2.11. The outputs c or d of $\mathcal{B}$ are indicated inside the states. The Cartesian product $\mathcal{P}$ of these two automata[7] is depicted on the right in the lower part of Figure 2.11. The ANS $\mathcal{S}$ and the DFAO $\mathcal{B}$ define the $\mathcal{S}$-automatic sequence

x = ccdcdccdcccdccccdcccccdccccccd····.

It is easy to check that the $n$ht and $(n + 1)$st occurrences of d are separated by $n$ letters $c$'s.

With the notation of the previous proof, we define a morphism

$$\psi_{\mathcal{P}} : z \mapsto z1,\ 1 \mapsto 12,\ 2 \mapsto 43,\ 3 \mapsto 63,\ 4 \mapsto 45,\ 5 \mapsto 46,\ 6 \mapsto 66,$$

and a coding

$$\nu : z, 4, 5, 6 \mapsto \varepsilon,\ 1, 3 \mapsto c,\ 2 \mapsto d.$$

---

[7] It is not the first time that we see this automaton. See example 1.11.

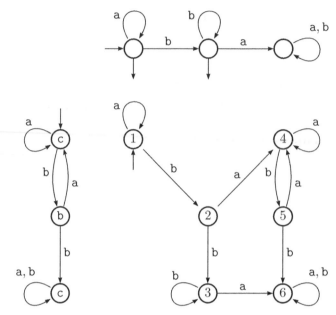

**Figure 2.11.** *Illustrating the proof of proposition 2.32*

Since $\{4, 5, 6\}$ is a strongly connected component of $\mathcal{P}$, we can directly replace $\psi_{\mathcal{P}}$ with $\psi'_{\mathcal{P}}$ defined by

$$\psi'_{\mathcal{P}} : z \mapsto z1,\ 1 \mapsto 12,\ 2 \mapsto 3,\ 3 \mapsto 3.$$

The $\mathcal{S}$-automatic word x is equal to $\nu(\psi_{\mathcal{P}}'^{\omega}(z))$.

### 2.2.2. *Some properties of abstract numeration systems*

We quickly summarize some of the research directions that were considered about ANS. We introduced the notion of $k$-recognizable set of integers in definition 1.35. Since we have a generalization of the classical integer base system, the following definition is quite natural. The motivation is to study sets of numbers whose language of representations within a fixed numeration system have a simple expression. From that point of view regular languages are accepted by the simplest machines and, thus, we will consider sets of numbers whose representations form a regular language.

DEFINITION 2.34.– Let $S$ be an ANS. A set $X \subseteq \mathbb{N}$ is *S-recognizable* if $\operatorname{rep}_S(X)$ is a regular language.

Theorem 1.33 can be restated as follows. Let $S$ be an ANS. Since the set of regular languages is closed under (finite) union, every ultimately periodic set is $S$-recognizable. Hence, every divisibility criterion can again be checked by means of finite automata when integers are represented within an ANS.

We have a direct analogue of proposition 1.37.

PROPOSITION 2.35.– Let $S$ be an ANS. A set $X \subseteq \mathbb{N}$ is $S$-recognizable if and only if its characteristic sequence is $S$-automatic.

The notion of kernel of a sequence can be generalized as follows. Recall that to get an alternative presentation of the $k$-kernel of a sequence (see definition 2.45, Volume 1), we have introduced an operator

$$\operatorname{sub}_k : [\![0, k-1]\!]^* \times B^{\mathbb{N}} \to B^{\mathbb{N}}$$

where $\operatorname{sub}_k(s, \mathbf{x})$ is the subsequence obtained by conserving in $\mathbf{x} \in B^{\mathbb{N}}$ its $j$th term if and only if the base-$k$ expansion of $j$ ends with $s$. Let $S$ be an ANS over a regular language $L \subseteq A^*$. We extend this operator to

$$\operatorname{sub}_S : A^* \times B^{\mathbb{N}} \to B^{\mathbb{N}}$$

where $\operatorname{sub}_S(s, \mathbf{x})$ is the subsequence obtained by conserving in $\mathbf{x} \in B^{\mathbb{N}}$ its $j$th term if and only if $\operatorname{rep}_S(j)$ ends with $s$.

Note that $\mathrm{sub}_S(s, \mathbf{x})$ can be finite or empty if the set of $L \cap A^*s$ is finite or empty, respectively. The *S-kernel* of an infinite word $\mathbf{x} \in B^{\mathbb{N}}$ is the set

$$\{\mathrm{sub}_S(s, \mathbf{x}) \mid s \in A^*\}.$$

PROPOSITION 2.36.– [RIG 02b] An infinite word $\mathbf{x} \in B^{\mathbb{N}}$ is *S*-automatic if and only if its *S*-kernel is finite.

Let $S = (L, A, <)$ be an ANS. The set of *S*-automatic sequences is closed under finite modifications. This is the natural analogue of propositions 2.38 and 2.39, Volume 1.

Recall that if $\mathbf{x} = (x_n)_{n \geq 0}$ is a $k$-automatic word, then the periodic decimation $(x_{an+b})_{n \geq 0}$ is again $k$-automatic; see proposition 2.36, Volume 1. In this more general setting, the situation is a bit more complicated.

PROPOSITION 2.37.– [RIG 02b] Let $S = (L, A, <)$ be an ANS and x be an *S*-automatic word. If the sequence y is a periodic decimation of x, then there exists an ANS $\mathcal{T}$ such that y is $\mathcal{T}$-automatic. Moreover, the set of *S*-automatic sequences is not closed under periodic decimation, i.e. there exist an *S*-automatic word $\mathbf{x} = (x_n)_{n \geq 0}$ and integers $a, b$ such that $(x_{an+b})_{n \geq 0}$ is not *S*-automatic.

In particular, this means that periodic decimation of morphic words are morphic.

In this setting of ANS, there are several directions of research. For a given ANS $S$, we could try to characterize the *S*-recognizable sets. Conversely, for a given set of integers, we could try to find an ANS $S$ for which this set is *S*-recognizable. See, for instance, Chapter 3 in [BER 10]. For a generalization to a multidimensional setting, see, for instance, [CHA 10].

For the representation of real numbers using ANS, there is a short description at the end of Chapter 3 in [BER 10]. Also see [LEC 02, LEC 04].

## 2.3. Positional numeration systems

Let $U = (U_n)_{n \geq 0}$ be an increasing sequence of integers. We assume that $U_0 = 1$ to ensure that every positive integer has at least one representation. We compute at first what is called the greedy representation of an integer.

Let $n$ be a positive integer. Find the largest $\ell$ such that $U_{\ell-1} \leq n < U_\ell$. Consider the Euclidean division of $n$ by $U_{\ell-1}$. Hence, $n = c_{\ell-1} U_{\ell-1} + r_{\ell-1}$ with $r_{\ell-1} < U_{\ell-1}$. Repeat the procedure with the remainder to get

$$n = \sum_{j=0}^{\ell-1} c_j\, U_j \,,$$

where the $c_j$'s are non-negative integers and $c_{\ell-1}$ is non-zero. Note that, for all $j \in [\![0, \ell - 1]\!]$, we have

$$c_j \leq \left\lfloor \frac{U_{j+1} - 1}{U_j} \right\rfloor .$$

The procedure that we have used is greedy, meaning that, for all $t \in [\![0, \ell - 1]\!]$,

$$\sum_{j=0}^{t} c_j\, U_j < U_{t+1} . \tag{2.9}$$

DEFINITION 2.38.– The word $c_{\ell-1} \cdots c_0$ is usually called the *normal U-representation* of $n$ and is denoted by $\mathrm{rep}_U(n)$. The empty word is representing 0. We refer to $\mathrm{rep}_U(\mathbb{N})$ as *the language of the numeration*.

DEFINITION 2.39.– If $d_r \cdots d_0$ is a word over $\mathbb{Z}$, then we define its *(U-)numerical value*

$$\mathrm{val}_U(d_r \cdots d_0) = \sum_{i=0}^{r} d_i \, U_i \, .$$

If $\mathrm{val}_U(d_r \cdots d_0) = n$, then the word $d_r \cdots d_0$ is a *U-representation* of $n$ (but not necessarily the normal $U$-representation).

To ensure that the normal $U$-representation of every integer is expressed over some finite alphabet, we will, moreover, assume that $\sup_n \frac{U_{n+1}}{U_n}$ is bounded by a constant $C$. Indeed, due to the greediness of the decomposition, if $U_{j+1} \leq C U_j$, then in a normal $U$-representation $c_{\ell-1} \cdots c_j \cdots c_0$ the coefficient $c_j$ is less than $C$.

We let $A_U$ denote the minimal alphabet such that every word in $\mathrm{rep}_U(\mathbb{N})$ is expressed over $A_U$.

EXAMPLE 2.40.– As an example, consider the Fibonacci sequence $(F_n)_{n\geq 0}$ defined by $F_{n+2} = F_{n+1} + F_n$, for all $n \geq 0$, and starting[8] with $F_0 = 1$ and $F_1 = 2$. This system is also referred to as the *Zeckendorf[9] numeration system* [ZEC 72]. For instance, in this system, $A_F = \{0, 1\}$ and 10 has a normal $F$-representation given by 10010 but 1110 is also a $F$-representation of 10 because $10 = 5 + 3 + 2$.

---

8 The reader may have noticed that we conveniently use different indexing of the Fibonacci sequence in different parts of the book. Indeed, when discussing numeration systems, it is quite convenient to start with a first term equal to one and use an increasing sequence.

9 Édouard Zeckendorf (1901–1983) was a Belgian physician serving as an army officer in the Belgian army. He graduated as medical doctor from the University of Liège [KIM 98]. In his paper [ZEC 72], he explains that his result about the decomposition of integers with non-consecutive Fibonacci numbers was obtained in 1939 and that J.F. Koksma asked C.G. Lekkerkerker to write a proof of his result.

The systems we are dealing with in this section are called *positional numeration systems* (PNS). This means that there exists an increasing sequence $U = (U_n)_{n \geq 0}$ such that we can make use of the map $\mathrm{val}_U : \mathbb{Z}^* \to \mathbb{Z}$ given in definition 2.39 and, for every integer $n$, if $w_t \cdots w_0$ is the canonical[10] expansion of $n$ within the considered system, then $n = \mathrm{val}_U(w_t \cdots w_0)$.

REMARK 2.41.– The reader may wonder what are the differences or advantages of using PNSs instead of ANS. For a given PNS $U = (U_n)_{n \geq 0}$, if $\mathrm{rep}_U(n) = c_{\ell-1} \cdots c_0$ and $\mathrm{rep}_U(m) = d_{k-1} \cdots d_0$ with $k \leq \ell$, then we trivially get the following decomposition

$$m + n = \sum_{j=0}^{\ell-1} (c_j + d_j)\, U_j \qquad\qquad [2.10]$$

by setting $d_i = 0$ for all $i \geq k$. Of course, such a decomposition does not necessarily fulfill the greediness condition [2.9], but we still get some information about $m + n$ from the $U$-representations of $m$ and $n$. In general, such a reasoning cannot be achieved for ANS because the value of a word depends only on its position within a language (see remark 2.43). Therefore, having a sequence $(U_n)_{n \geq 0}$ associated with a positional system can be useful when dealing with arithmetic operations. It permits writing expressions like [2.10].

In contrast to the ANS $\mathcal{S}$ given in example 2.18, we have $\mathrm{rep}_{\mathcal{S}}(3) = $ aa and $\mathrm{rep}_{\mathcal{S}}(7) = $ aab. But this knowledge does not give any information about $\mathrm{rep}_{\mathcal{S}}(10)$, which turns out to be aaaa.

LEMMA 2.42.– Let $U = (U_n)_{n \geq 0}$ be an increasing sequence of integers such that $U_0 = 1$. If $d_r \cdots d_0 \in \mathbb{N}^*$ is a

---

10 We just mean that a particular expansion is associated with every integer.

$U$-representation of $n \geq 1$ with $d_r \neq 0$, i.e. $\mathrm{val}_U(d_r \cdots d_0) = n$, then the word $d_r \cdots d_0$ is genealogically less than or equal to the normal $U$-representation $\mathrm{rep}_U(n)$. Moreover, $n < m$ if and only if $\mathrm{rep}_U(n)$ is genealogically less than $\mathrm{rep}_U(m)$.

PROOF.– It follows directly from the greediness of the algorithm computing the normal $U$-representation of $n$.  ∎

As a result of the second part of lemma 2.42, if a PNS $U$ is such that $\mathrm{rep}_U(\mathbb{N})$ is a regular language, then it coincides with an ANS. In Figure 2.12, $\mathcal{A}$ is the set of ANS for which there is no equivalent[11] PNS constructed on an increasing sequence $(U_n)_{n \geq 0}$. This set is non-empty as explained in the following remark 2.43. The set $\mathcal{B}$ contains all the PNSs $U$ such that the language of the numeration $\mathrm{rep}_U(\mathbb{N})$ is regular. In particular, $\mathcal{B}$ contains the Pisot systems discussed in the next section[12]. Finally, $\mathcal{C}$ contains every positional system $U$ such that $\mathrm{rep}_U(\mathbb{N})$ is not regular (for instance, if $U$ does not satisfy any linear recurrence relation, then we can use proposition 2.46; or if $\sup_n \frac{U_{n+1}}{U_n}$ is unbounded, then the alphabet $A_U$ is infinite).

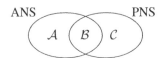

**Figure 2.12.** *Abstract numeration systems and positional numeration systems*

REMARK 2.43.– ANS do not always define a PNS. Let us pursue example 2.18 about $\mathcal{S} = (\mathrm{a}^*\mathrm{b}^*, \{\mathrm{a}, \mathrm{b}\}, \mathrm{a} < \mathrm{b})$ a little bit further. Assume that we have a map $v : \{\mathrm{a}, \mathrm{b}\} \to \mathbb{N}$ which

---

11 Two systems are equivalent if they provide the same representations up to a coding of the digit set.

12 But the sequence $(U_n)_{n \geq 0}$ defined by $U_0 = 1$, $U_1 = 3$ and $U_{n+2} = U_{n+1} + 2U_n$ is such that $0^* \mathrm{rep}_U(\mathbb{N}) = \{0, 1\}^* \cup \{0, 1\}^* 02(00)^*$. This latter example is taken from [FRO 96a]. Some more examples can be found in [FRO 97].

assigns some weight to a and b. We show that there exists *no* sequence $U = (U_n)_{n \geq 0}$ defining a PNS such that, for all words $w_\ell \cdots w_0 \in L$,

$$\mathrm{val}_S(w_\ell \cdots w_0) = \sum_{k=0}^{\ell} v(w_k) U_k.$$

We proceed by contradiction and we assume that such a sequence exists. Since $U_0 = 1$ and $\mathrm{val}_S(\mathrm{a}) = 1$, $\mathrm{val}_S(\mathrm{b}) = 2$, we must have $v(\mathrm{a}) = 1$ and $v(\mathrm{b}) = 2$. Notice that $\mathrm{val}_S(\mathrm{aa}) = 3$ and this quantity should be equal to $v(\mathrm{a})U_1 + v(\mathrm{a})U_0$. Consequently, $U_1 = 2$. Therefore, $v(\mathrm{b})U_1 + v(\mathrm{b})U_0 = 6$ but $\mathrm{val}_S(\mathrm{bb}) = 5$, which is contradictory.

DEFINITION 2.44.– A natural framework is to consider PNSs constructed on a sequence $U = (U_n)_{n \geq 0}$ of integers satisfying a linear recurrence relation: there exist $a_0, \ldots, a_{k-1} \in \mathbb{Z}$ such that

$$\forall n \geq 0, \ U_{n+k} = a_{k-1} U_{n+k-1} + \cdots + a_0 U_n. \qquad [2.11]$$

Recall that we assume that the sequence is increasing, $U_0 = 1$ and $\sup_n \frac{U_{n+1}}{U_n}$ is bounded by a constant. In this case, we say that $U$ is a *linear numeration system*.

If $\lim_{n \to +\infty} U_{n+1}/U_n = \beta$ for some real $\beta > 1$, then $U$ is said to *satisfy the dominant root condition* and $\beta$ is called the *dominant root* of the recurrence. Such a condition is not always satisfied. For instance, take the sequence $(V_n)_{n \geq 0}$ where $V_0 = 1$, $V_{2n+1} = 2V_{2n}$ and $V_{2n+2} = 3V_{2n+1}$ for all $n \geq 0$, $(V_n)_{n \geq 0} = 1, 2, 6, 12, 36, 72, 216, \ldots$. Such a sequence does not satisfy the dominant root condition. Nevertheless, a sequence such as the Fibonacci sequence (and more generally, the sequences associated with Pisot numbers discussed in the next section) satisfies the dominant root condition.

So far we have introduced $k$-recognizable sets and $S$-recognizable sets of integers (for integer base systems and

ANS), see definitions 1.35 and 2.34, respectively. The same definition is adapted to linear numeration systems.

DEFINITION 2.45.– Let $U$ be a linear numeration system. A set $X \subseteq \mathbb{N}$ is said to be *U-recognizable* if $\mathrm{rep}_U(X)$ is a regular language[13]. We can also define $U$-recognizable subsets of $\mathbb{N}^d$. Let $X^d \subseteq \mathbb{N}$. We say that $X$ is *U-recognizable* if the language

$$\mathrm{rep}_U(X) = \left\{ \left( 0^{\ell - |\mathrm{rep}_U(x_1)|} \mathrm{rep}_U(x_1), \dots, 0^{\ell - |\mathrm{rep}_U(x_d)|} \mathrm{rep}_U(x_d) \right) \right.$$

$$\left. \mid (x_1, \dots, x_d) \in X,\ \ell = \max_i |\mathrm{rep}_U(x_i)| \right\}$$

is accepted by a DFA over the alphabet $(A_U)^d$ as considered in section 1.3.

PROPOSITION 2.46.– Let $U = (U_n)_{n \geq 0}$ be an increasing sequence of integers such that $U_0 = 1$ and $\sup_n \frac{U_{n+1}}{U_n}$ is bounded by a constant. If the language of numeration is regular, i.e. if $\mathrm{rep}_U(\mathbb{N})$ is accepted by some DFA, then $U$ is a linear numeration system.

PROOF.– Note that $\mathrm{rep}_U(U_\ell) = 10^\ell$ for all $\ell \geq 0$. Among the words of length $\ell + 1$ in $\mathrm{rep}_U(\mathbb{N})$, the smallest one for the genealogical order is $10^\ell$. Consequently, for all $\ell \geq 0$, $U_{\ell+1} - U_\ell$ is exactly the number of words of length $\ell + 1$ in $\mathrm{rep}_U(\mathbb{N})$. Since the latter language is regular, it is accepted by a DFA and the number of words of length $n$ in $\mathrm{rep}_U(\mathbb{N})$ is equal to the number of paths of length $n$ from the initial state to the final ones. From section 1.2, we deduce that the sequence $(U_n - U_{n-1})_{n \geq 1} = (\mathrm{Card}(\mathrm{rep}_U(\mathbb{N}) \cap A_U^n))_{n \geq 1}$ satisfies a linear recurrence relation with integer coefficients and the conclusion about $(U_n)_{n \geq 0}$ follows easily.  ∎

_____

13 Note that $\mathrm{rep}_U(X)$ is a regular language if and only if $0^* \mathrm{rep}_U(X)$ is regular.

EXERCISE 2.3.1.– [SHA 94] Let $k, d$ be positive integers such that $d \geq k$. Let $M$ be a $k \times d$ matrix with integer entries. Let $(A_n)_{n\geq0}$ be a sequence such that, for all sufficiently large $n$, we have

$$
\begin{cases}
A_{kn} = M_{11}A_{kn-1} + M_{12}A_{kn-2} + \cdots + M_{1d}A_{kn-d} \\
A_{kn-1} = M_{21}A_{kn-2} + M_{22}A_{kn-3} + \cdots + M_{2d}A_{kn-d-1} \\
\quad \vdots \\
A_{kn-k+1} = M_{k1}A_{kn-k} + M_{k2}A_{kn-k-1} + \cdots + M_{kd}A_{kn-k-d+1}.
\end{cases}
$$

The sequence $(A_n)_{n\geq0}$ satisfies a linear recurrence relation with constant coefficients.

Note that considering a linear numeration system does not imply, in general, that $\mathbb{N}$ is $U$-recognizable.

EXAMPLE 2.47.– [SHA 94] Such a counterexample is given by the sequence $(U_n)_{n\geq0}$ defined by $U_n = (n+1)^2$. We have $U_0 = 1$, $U_1 = 4$, $U_2 = 9$ and $U_{n+3} = 3U_{n+2} - 3U_{n+1} + U_n$. In that case, $\operatorname{rep}_U(\mathbb{N}) \cap 10^*10^* = \{10^a10^b \mid b^2 < 2a + 4\}$ showing with the pumping lemma, lemma 1.20, that $\mathbb{N}$ is not $U$-recognizable.

THEOREM 2.48 (Folklore).– [BER 10, Prop. 3.1.9] Let $p, r \geq 0$. If $U = (U_n)_{n\geq0}$ is a linear numeration system, then

$$
\left\{ c_\ell \cdots c_0 \in A_U^* \mid \sum_{k=0}^{\ell} c_k U_k \in p\mathbb{N} + r \right\}
$$

is accepted by a DFA that can be effectively constructed. In particular, if $\mathbb{N}$ is $U$-recognizable, then every ultimately periodic set is $U$-recognizable.

For the particular case, we use the fact that the set of regular languages is closed under intersection.

EXAMPLE 2.49.– Consider the Fibonacci numeration system given by the sequence $F_0 = 1$, $F_1 = 2$ and $F_{n+2} = F_{n+1} + F_n$ for all $n \geq 0$. For this system, $0^* \operatorname{rep}_F(\mathbb{N})$ is given by the set of words over $\{0,1\}$ avoiding the factor $11$ and the set of even numbers is $F$-recognizable [CHA 11] using the DFA shown in Figure 2.13.

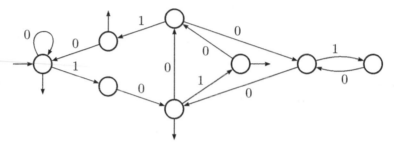

**Figure 2.13.** *A (trim) DFA accepting* $0^* \operatorname{rep}_F(2\mathbb{N})$

| 0 | $\varepsilon$ | 10 | 10010 | 20 | 101010 |
|---|---|---|---|---|---|
| 2 | 10 | 12 | 10101 | 22 | 1000001 |
| 4 | 101 | 14 | 100001 | 24 | 1000100 |
| 6 | 1001 | 16 | 100100 | 26 | 1001000 |
| 8 | 10000 | 18 | 101000 | 28 | 1001010 |

**Table 2.4.** *The Zeckendorf expansions of the first few even integers*

EXERCISE 2.3.2.– Use the fact that the Fibonacci sequence modulo 2 is periodic of period 3: $1, 0, 1, 1, 0, 1, \ldots$ and build a DFA recognizing $0^* \operatorname{rep}_F(2\mathbb{N})$ when reading $F$-representations least significant digits first. From this DFA derive the DFA depicted in Figure 2.13.

To finish this section, we show that $U$-recognizable sets are recognized by some automata having a special form associated with the minimal automaton of $0^* \operatorname{rep}_U(\mathbb{N})$.

Let $U$ be a PNS such that $\mathbb{N}$ is $U$-recognizable. Let $\mathcal{N}_U = (Q_U, q_{0,U}, A_U, \delta_U, F_U)$ be the minimal automaton of $0^* \operatorname{rep}_U(\mathbb{N})$. We will consider conditions coming from automata theory similar to those of theorem 1.26. Note that the following definition and statements can be extended to subsets of $\mathbb{N}^d$. See, for instance, [BRU 97].

DEFINITION 2.50.– Let $\mathcal{A} = (Q, q_0, A_U, \delta, F)$ be a DFA. This DFA is said to be a $U$-*automaton* if there exists a map $\Phi : Q \to Q_U$ (i.e. a morphism of automata) such that

- $\Phi$ is onto;
- $\Phi(q_0) = q_{0,U}$;
- for all $c \in A_U$ and $q \in Q$, $\Phi(\delta(q, c)) = \delta_U(\Phi(q), c)$.

Moreover, there exists $F'$ such that $F \subseteq F' \subseteq Q$[14]:

- $\Phi(F') = F_U$;
- $\Phi^{-1}(F_U) = F'$.

PROPOSITION 2.51.– Let $U$ be a PNS such that $\mathbb{N}$ is $U$-recognizable. An accessible DFA $\mathcal{A} = (Q, q_0, A_U, \delta, F)$ is a $U$-automaton if and only if there exists $F'$ such that $F \subseteq F' \subseteq Q$ and $\mathcal{A}' = (Q, q_0, A_U, \delta, F')$ accepts $0^* \operatorname{rep}_U(\mathbb{N})$.

PROOF.– Assume that $\mathcal{A}$ is a $U$-automaton. The conclusion follows directly from the morphism $\Phi$ and the set $F'$ given in the definition.

For the converse, assume that there exists $F'$ such that $F \subseteq F' \subseteq Q$ and $\mathcal{A}' = (Q, q_0, A_U, \delta, F')$ accepts $0^* \operatorname{rep}_U(\mathbb{N})$. Then from theorem 1.26, there exists a morphism $\Phi$ between $\mathcal{A}'$ and minimal automaton $\mathcal{N}_U$ accepting the same language. ∎

PROPOSITION 2.52.– Let $U$ be a PNS such that $\mathbb{N}$ is $U$-recognizable. A set $X \subset \mathbb{N}$ is $U$-recognizable if and only if the set $0^* \operatorname{rep}_U(X)$ is accepted by a $U$-automaton.

PROOF.– If $0^* \operatorname{rep}_U(X)$ is accepted by a $U$-automaton, then in particular, $X$ is $U$-recognizable.

---

14 Note that these conditions are replacing the condition $\Phi(F) = F_U$ given in theorem 1.26. In particular, $\mathcal{A}$ recognizes a subset of $0^* \operatorname{rep}_U(\mathbb{N})$.

Assume that $0^* \mathrm{rep}_U(X)$ is accepted by some DFA $\mathcal{A} = (Q, q_0, A_U, \delta, F)$. Consider the product, as given in definition 1.10, of the two automata $\mathcal{A}$ and $\mathcal{N}_U$. Assuming that a state $(q, q')$ in the product is final if and only if $q \in F$ implies that this product automaton accepts $0^* \mathrm{rep}_U(X)$. Note that if $(q, q')$ is an accessible state with $q \in F$, then $q'$ must be a final state in $\mathcal{N}_U$ because $\mathcal{A}$ accepts a sublanguage of the language accepted by $\mathcal{N}_U$. Projection on the second component is clearly a morphism of automata between $\mathcal{A} \times \mathcal{N}_U$ and $\mathcal{N}_U$. Finally, the last condition about $F'$ is fulfilled by taking $F'$ to be the set of pairs $(q, q')$ where $q'$ is final.   ∎

EXAMPLE 2.53.– In Figure 2.14 we have depicted the minimal automaton corresponding to the Zeckendorf expansions (below). In the upper part, we have considered a DFA accepting $U$-representations containing an even number of 1's. The first few integers in the corresponding $U$-recognizable set are $0, 1, 4, 6, 9$. To see that the DFA is indeed a $U$-automaton, consider the map $\Phi : q, s \mapsto q_0,\ r, t \mapsto q_1$. With the notation of definition 2.14, take $F' = \{q, r, s, t\}$.

## 2.4. Pisot numeration systems

We now consider particular linear numeration systems having rich properties (the language of the numeration is regular, addition can be performed by a DFA, etc). As we will see in the next and last chapter of this book, these properties are so strong that each set of integers recognizable within such a system is characterized by some first-order formula in the formalism of mathematical logic. This family of systems contains integer base systems and the Fibonacci or tribonacci systems.

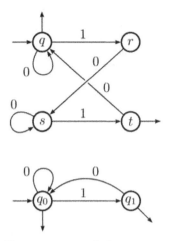

**Figure 2.14.** *A U-automaton and the corresponding minimal automaton $\mathcal{N}_U$*

DEFINITION 2.54.– Let $U = (U_n)_{n\geq 0}$ be a linear numeration system satisfying equation [2.11]. Consider the characteristic polynomial of the recurrence given by

$$P(X) = X^k - a_{k-1}X^{k-1} - \cdots - a_1 X - a_0.$$

In what follows, we assume that $P$ is the minimal polynomial of a Pisot number $\alpha > 1$. Recall that this means that the other roots of $P$ have modulus less than one. In that case, we will say that $U = (U_n)_{n\geq 0}$ is a *Pisot numeration system*.

Note that if two Pisot numeration systems are associated with the same Pisot number, then these two systems differ only by the choice of the initial values $U_0, \ldots, U_{k-1}$.

Since $\alpha = \alpha_1$ is a Pisot number (see section 1.1.2, Volume 1), the polynomial $P$ can be factored as

$$P(X) = (X - \alpha_1) \cdots (X - \alpha_k)$$

where the complex numbers $\alpha_1, \ldots, \alpha_k$ are pairwise distinct[15] and, for all $j > 1$, we have $|\alpha_j| < 1$.

Using theorem 1.16, we get

$$U_n = c_1 \alpha_1^n + c_2 \alpha_2^n + \cdots + c_k \alpha_k^n \quad \forall n \geq 0$$

and

$$\lim_{n \to +\infty} \frac{U_n}{\alpha_1^n} = c_1$$

because, for all $j > 1$, $|\alpha_j^n| \to 0$ if $n \to +\infty$. In particular, we deduce that $U_{n+1}/U_n$ tends to $\alpha_1$ and, thus, Pisot systems satisfy the dominant root condition.

Pisot numeration systems have nice properties that we now list. For a proof of the following result, see [BRU 97, Theorem 8] or [FRO 96b].

PROPOSITION 2.55.– Let $U = (U_n)_{n \geq 0}$ be a Pisot numeration system. The languages $\mathrm{rep}_U(\mathbb{N})$ and $0^* \, \mathrm{rep}_U(\mathbb{N})$ are regular, i.e. $\mathbb{N}$ is $U$-recognizable.

REMARK 2.56.– From this result and lemma 2.42, we know that every Pisot numeration system is a special case of an ANS.

EXAMPLE 2.57.– The usual integer base system is a special case of a Pisot system. Indeed, the only rational numbers that are Pisot numbers are the integers greater than one.

EXAMPLE 2.58 (Modified Fibonacci system).– Consider the sequence $U = (U_n)_{n \geq 0}$ defined by the recurrence $U_{n+2} = U_{n+1} + U_n$ but with the initial conditions $U_0 = 1$, $U_1 = 3$. We get a numeration system

---

15 If $M_\alpha$ is the minimal polynomial of $\alpha$, then $\alpha$ is a simple root of $M_\alpha$. If $\beta$ is a root of $M_\alpha$, then $M_\alpha$ is also the minimal polynomial of $\beta$.

$(U_n)_{n \geq 0} = 1, 3, 4, 7, 11, 18, 29, 47, \ldots$ constructed from the Lucas numbers. The $U$-representations of the first few integers within this system are

$\varepsilon$, 1, 2, 10, 100, 101, 102, 1000, 1001, 1002, 1010, 10000, $\ldots$ .

So except for the least significant digit, these $U$-representations share the same property as the representations in the Zeckendorf system: the factor 11 is forbidden and we use only the digits 0 and 1. For this system, $A_U = \{0, 1, 2\}$ and the minimal automaton of $0^* \operatorname{rep}_U(\mathbb{N})$ is depicted in Figure 2.15. For details, see [CHA 11].

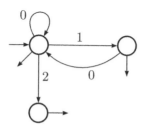

**Figure 2.15.** *The DFA recognizing* $0^* \operatorname{rep}_U(\mathbb{N})$ *for the modified Fibonacci system*

If instead we consider the same recurrence relation but with the usual initial conditions $U_0 = 1$ and $U_1 = 2$, then $0^* \operatorname{rep}_U(\mathbb{N}) = \{0, 1\}^* \setminus \{0, 1\}^* 11 \{0, 1\}^*$.

Since these two systems are associated with the same recurrence relation, they are two Pisot numeration systems associated with the golden ratio $(1 + \sqrt{5})/2$ whose minimal polynomial is $X^2 - X - 1$.

DEFINITION 2.59.– Let $B \subset \mathbb{Z}$ be a finite alphabet. Let $w_r \cdots w_0$ be a finite word over $B$ such that $\operatorname{val}_U(w_r \cdots w_0) \geq 0$. We add this latter condition because, using negative digits, a word might have a negative $U$-numerical value. Then we can consider the *normalization map* sending $w_r \cdots w_0$ to $\operatorname{rep}_U(\operatorname{val}_U(w_r \cdots w_0))$. This map is denoted by

$$\nu_{U,B} : v \in \{w \in B^* \mid \operatorname{val}_U(w) \geq 0\} \mapsto \operatorname{rep}_U(\operatorname{val}_U(v)).$$

The following result is fundamental for Pisot systems. For a proof, see [FRO 92a] or [BRU 97, Theorem 14]. In the statement, we use definition 2.45 for subsets of $\mathbb{N}^2$ and multidimensional alphabets as in section 1.3.

THEOREM 2.60 (Frougny's normalization).– Let $B \subset \mathbb{Z}$ be a finite alphabet. Let $U = (U_n)_{n \geq 0}$ be a Pisot numeration system. The graph of the normalization map $\nu_{U,B}$ is $U$-recognizable, i.e.

$$\left\{ \left( 0^{\ell - |w|} w, 0^{\ell - |\nu_{U,B}(w)|} \nu_{U,B}(w) \right) \mid w \in B^*, \mathrm{val}_U(w) \geq 0 \right\}$$
$$\subseteq (B \times A_U)^*$$

is accepted by a DFA.

EXAMPLE 2.61.– We consider base-2 representations written, respectively, over the alphabets $\{0,1,2\}$, $\{0,1,2,3\}$ and $\{-1,0,1\}$. We write $\bar{1}$ instead of $-1$. The three automata depicted in Figure 2.16 are constructed on the same model. We read representations' least significant digit first. Each state is used to store the current carry. Note that all edges are of the form

$$c \xrightarrow{\binom{a}{b}} c'$$

where $b \in \{0,1\}$ and

$$c + a - b = 2c'. \tag{2.12}$$

Roughly speaking, reading the digit $a$ and taking into account the carry $c$, we get as output a digit $b$ and we are left with a carry for the next position.

We have seen in example 1.19 that for base-2 expansions we can produce a DFA mimicking addition. Actually this can be done for every Pisot system.

COROLLARY 2.62.– Let $U = (U_n)_{n \geq 0}$ be a Pisot numeration system. Addition and subtraction in $\mathbb{N}$ are $U$-recognizable, i.e. the sets $\{(x, y, z) \in \mathbb{N}^3 \mid x+y = z\}$ and $\{(x, y, z) \in \mathbb{N}^3 \mid x-y = z\}$ are $U$-recognizable.

The idea is to perform addition (or subtraction) component-wise and then apply normalization. Let $A_U = \{0, \ldots, c\}$. Then for addition (respectively, subtraction), consider normalization over the alphabet $A_U + A_U = \{0, \ldots, 2c\}$ (respectively, the symmetrized alphabet $A_U - A_U = \{-c, \ldots, c\}$).

PROOF.– Let $B = \{0, \ldots, 2c\}$. Consider the language

$$\text{Add} = \{(a, b, a + b, d) \mid a, b, d \in A_U\}^*$$

over the alphabet $A_U \times A_U \times B \times A_U$. This language is clearly regular. For the second part of the statement, we can adapt this proof to the language

$$\{(a, b, a - b, d) \mid a, b, d \in A_U, \ a \geq b\}^*$$

Consider the morphism

$$p : (A_U \times A_U \times B \times A_U)^* \rightarrow (B \times A_U)^*,$$
$$(x_1, x_2, x_3, x_4) \mapsto (x_3, x_4).$$

From the normalization theorem of Frougny, we know that the graph of $\nu_{U,B}$ in $\mathbb{N}^2$ is $U$-recognizable. So the corresponding set of $U$-representations is a regular language $N$ over $B \times A_U$. From proposition 1.30, the language $p^{-1}(N)$ is regular and thus $p^{-1}(N) \cap \text{Add}$ is regular. Now apply the morphism $q : (x_1, x_2, x_3, x_4) \mapsto (x_1, x_2, x_4)$ to this language to get the expected language. ∎

COROLLARY 2.63.– Let $U = (U_n)_{n \geq 0}$ be a Pisot numeration system. Multiplication by a constant $c \in \mathbb{N}$ is $U$-recognizable, i.e. the set $\{(x, cx) \in \mathbb{N}^2 \mid x \geq 0\}$ is $U$-recognizable.

PROOF.– Left as an exercise. Adapt the previous proof. ∎

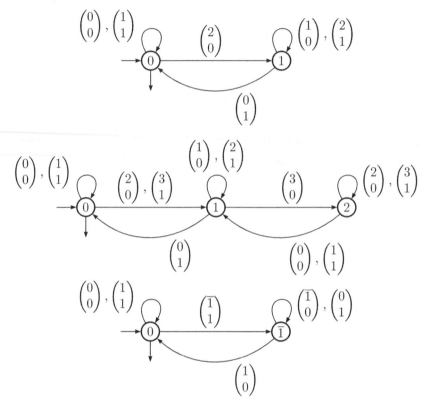

**Figure 2.16.** *Normalization in base 2 for the alphabets $\{0,1,2\}$,*
*$\{0,1,2,3\}$ and $\{-1,0,1\}$*

EXERCISE 2.4.1.– Consider base-2 expansions. Construct a
DFA reading least significant digit first and recognizing the
set of pairs $(x, 5x)$ written in base 2. The idea is to construct a
DFA for the normalization of representations over the
alphabet $\{0,5\}$. Indeed, if $c_\ell \cdots c_0$ is the base-2 expansion of $n$,
then $(5c_\ell)\cdots(5c_0)$ is a representation over $\{0,5\}$ of $5n$. It is
pretty easy to build such a normalizer using [2.12]. To get the
expected DFA, we have to replace every label of the form $\begin{pmatrix} 5 \\ a \end{pmatrix}$
with $\begin{pmatrix} 1 \\ a \end{pmatrix}$.

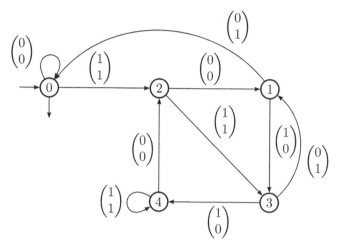

**Figure 2.17.** *Recognizing multiplication by* 5 *in base* 2

Extend this construction for base-$k$ expansions and multiplication by an arbitrary constant $c$. In particular, prove that the construction always provides a finite automaton.

REMARK 2.64.– There exists a non-Pisot system such that the language of the numeration is regular, but addition is not $U$-recognizable. See [FRO 96a, FRO 97, BEL 09].

DEFINITION 2.65.– A numeration system $U = (U_n)_{n \geq 0}$ is a *Bertrand numeration system* if, for all $w \in A_U^+$, $w \in \mathrm{rep}_U(\mathbb{N}) \Leftrightarrow w0 \in \mathrm{rep}_U(\mathbb{N})$.

As an example, the usual integer base systems and the Fibonacci or the tribonacci systems are Bertrand numeration systems. Details about the tribonacci system and connections to the tribonacci word are discussed in example 2.66.

For a given Pisot number $\alpha$, there exists a unique linear numeration system $U$ associated with $\alpha$ such that $U$ is also a Bertrand system. Of course, the linear recurrence relation is provided by the minimal polynomial of $\alpha$. As we will explain, the choice of the initial conditions can be derived from theorem 2.74. As an example, the Zeckendorf numeration system is a Bertrand system. But the modified version $(U_n)_{n\geq 0} = 1, 3, 4, 7, \ldots$ considered in example 2.58 is not a Bertrand system. Indeed, 2 is a greedy representation but not 20 because $\text{rep}_U(\text{val}_U(20)) = 102$. In the same way, for the sequence $(V_n)_{n\geq 0} = 1, 2, 3, 9, 27, \ldots$, the word 200 is a greedy representation (the normal $V$-representation of 6) but 20 and 2 are not greedy expansions ($\text{rep}_V(4) = 101$ and $\text{rep}_V(2) = 10$).

Recall that Pisot systems are special cases of ANS. Therefore, we can make use of the generalization of Cobham's theorem on automatic sequences given in theorem 2.24.

EXAMPLE    2.66.– Consider    the    tribonacci    sequence $T = (T_n)_{n\geq 0}$ defined by $T_0 = 1$, $T_1 = 2$, $T_2 = 4$ and $T_{n+3} = T_{n+2} + T_{n+1} + T_n$ for all $n \geq 0$. The linear numeration system $T$ is a Pisot system. Indeed, the characteristic polynomial of the relation is $X^3 - X^2 - X - 1$ and we have seen in section 2.2.2, Volume 1, that it is the minimal polynomial of a Pisot number.

From proposition 2.55, $0^* \text{rep}_T(\mathbb{N})$ is a regular language. It is easy to see that this language is $\{0, 1\}^* \setminus \{0, 1\}^* 111 \{0, 1\}^*$. It is accepted by the DFA depicted in Figure 2.18 where it is convenient that its states be denoted by $a$, $b$ and $c$. We can apply either lemma 2.25 or lemma 2.28 to this automaton. Since the DFA has a loop of label 0 on the initial state $a$, we choose to apply lemma 2.28. With this DFA and notation from lemma    2.28    we    associate    the    morphism $\mu : a \mapsto ab, b \mapsto ac, c \mapsto a$ which is exactly the tribonacci morphism. Also note that all states of the DFA are final. Hence, the language $M$ corresponding to all labels of paths that can be followed from the initial state, but not starting with 0, is exactly $\text{rep}_T(\mathbb{N})$. Lemma 2.28 states that $\mu^\omega(a) = x_0 x_1 x_2 \cdots$ (which is exactly the tribonacci word) is such that $x_n$ is the state reached when reading $\text{rep}_T(n)$. In particular, we get

$$
x_n = \begin{cases}
\text{a}, & \text{if } \text{rep}_T(n) \in \{\varepsilon\} \cup \{0,1\}^*0; \\
\text{b}, & \text{if } \text{rep}_T(n) \in \{1\} \cup \{0,1\}^*01; \\
\text{c}, & \text{if } \text{rep}_T(n) \in \{11\} \cup \{0,1\}^*011.
\end{cases}
$$

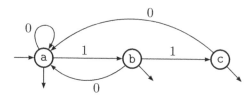

**Figure 2.18.** *The DFA recognizing* $0^* \text{rep}_T(\mathbb{N})$

## 2.5. Back to $\beta$-expansions

Even though Dumont–Thomas systems and ANS can be extended to allow the expansion of real numbers, for the first time in this chapter, we consider expansions of real numbers. We have already introduced $\beta$-expansions of real numbers in section 1.3.4, Volume 1.

### 2.5.1. *Representation of real numbers*

Let $x \in [0,1)$. Recall that $d_\beta(x)$ denotes the $\beta$-expansion of $x$ (which means that among all the $\beta$-representations of $x$ it is the representation obtained in a greedy way). We let $D_\beta$ denote the set of $\beta$-expansions of reals in $[0,1)$. The closure of $D_\beta$ is denoted by $S_\beta$ and is called the $\beta$-shift. See examples 1.66 and 1.67, Volume 1.

As shown by the following result referred to as Parry's theorem, the $\beta$-expansion of 1 is of particular importance. Let us start with a preliminary comment. So far we have defined $\beta$-expansions of real numbers in $[0,1)$. We define the

$\beta$-expansion of 1 as an infinite word over $[0, \lfloor\beta\rfloor - 1]$. In particular, this expansion is not $1 \bullet 000 \cdots$. The first digit of the $\beta$-expansion of 1 is set to $\lceil\beta\rceil - 1$. Then we have to consider the $\beta$-expansion of $1 - (\lceil\beta\rceil - 1)/\beta$ to get the other digits of the expansion, starting with the second digit. If $\beta$ belongs to $\mathbb{N}_{\geq 2}$, the $\beta$-expansion of 1 is $(\beta - 1)^{\omega}$.

For an example, the $\pi$-expansion of 1 starts with $c_0 \cdots c_7 = 30110211$ because the digits are in $[0, 3]$ and, for each $\ell \in [0, 7]$, we have

$$\sum_{j=0}^{\ell-1} \frac{c_j}{\pi^{j+1}} + \frac{c_\ell}{\pi^{\ell+1}} \leq 1 \quad \text{and} \quad \sum_{j=1}^{\ell-1} \frac{c_j}{\pi^{j+1}} + \frac{c_\ell + 1}{\pi^{\ell+1}} > 1.$$

Let $d_\beta(1) = (t_n)_{n \geq 1}$ be the $\beta$-expansion of 1. We define the *quasi-greedy expansion* $d_\beta^*(1)$ of 1 as follows. If $d_\beta(1) = t_1 \cdots t_m$ is finite, i.e. $t_m \neq 0$ and $t_j = 0$ for all $j > m$, then $d_\beta^*(1) = (t_1 \cdots t_{m-1}(t_m - 1))^{\omega}$, otherwise $d_\beta^*(1) = d_\beta(1)$. It is easy to see that

$$\text{if } 1 = \frac{t_1}{\beta} + \cdots + \frac{t_m}{\beta^m},$$

$$\text{then } 1 = \sum_{k=0}^{+\infty} \beta^{-km} \left( \frac{t_1}{\beta} + \cdots + \frac{t_{m-1}}{\beta^{m-1}} + \frac{t_m - 1}{\beta^m} \right).$$

An alternative definition of the quasi-greedy expansion of 1 is to set

$$d_\beta^*(1) = \lim_{x \to 1^-} d_\beta(x) \tag{2.13}$$

where the limit on the right-hand side is the limit of a sequence of infinite words.

EXAMPLE 2.67.– Consider the golden ratio $\varphi$. We have

$$1 = \frac{1}{\varphi} + \frac{1}{\varphi^2}.$$

Hence, $d_\beta(1) = 11$ (followed by $0^\omega$) and $d_\beta^*(1) = (10)^\omega$.

For the square of the golden ratio, we have

$$1 = \frac{2}{\varphi^2} + \sum_{i=2}^{+\infty} \frac{1}{(\varphi^2)^i}.$$

We can check that $d_\beta(1) = d_\beta^*(1) = 2\,1^\omega$.

Now that we have introduced the quasi-greedy expansion of 1, we can state Parry's theorem where the order used to compare infinite words is the usual lexicographic order (see definition 1.11, Volume 1).

THEOREM 2.68.– [PAR 60] Let $\beta > 1$ be a real number, and let s be an infinite sequence of non-negative integers. The sequence s belongs to $D_\beta$ if and only if, for all $k \geq 0$,

$$\sigma^k(s) < d_\beta^*(1)$$

and $s$ belongs to $S_\beta$ if and only if, for all $k \geq 0$,

$$\sigma^k(s) \leq d_\beta^*(1).$$

If $d_\beta^*(1)$ has no particular regularity, then the above theorem is of little practical interest. Nevertheless, we will see that Parry's theorem can be exploited whenever $d_\beta^*(1)$ is ultimately periodic because finite information is enough to completely encode such a word.

DEFINITION 2.69.– A real number $\beta$ such that $d_\beta(1)$ is eventually periodic is called a *Parry number*. If $d_\beta(1)$ is finite, i.e. $d_\beta(1)$ ends with $0^\omega$, then $\beta$ is called a *simple Parry number*. These numbers are sometimes called *-numbers* and *simple -numbers*.

An alternative definition for Parry numbers is to make use of the $\beta$-transformation $T_\beta : [0,1) \rightarrow [0,1)$ introduced in section 1.3.4, Volume 1. Assume that $\beta$ is not an integer. Note that the $\beta$-expansion of 1 is eventually periodic if $T^j(1 - (\lceil\beta\rceil - 1)/\beta) = T^k(1 - (\lceil\beta\rceil - 1)/\beta)$ for some $j < k$ or, equivalently, the set $\{T^n(1 - (\lceil\beta\rceil - 1)/\beta) \mid n \geq 1\}$ is finite.

EXAMPLE 2.70.– Continuing example 2.67, the golden ratio is a simple Parry number and its square is a non-simple Parry number. The fact that $d_\beta^*(1)$ is eventually periodic permits us to derive an automaton where the labels of the infinite paths that can be read are exactly the admissible infinite words devised by Parry's theorem. In Figure 2.19, the admissible paths are those that do not contain the factor 11. Indeed, assume that we consider an infinite word $w = x11z$ where $w, z \in \{0, 1\}^{\mathbb{N}}$ and $x \in \{0, 1\}^*$. Then the infinite word $\sigma^{|x|}(w)$ is lexicographically larger than $d_\varphi^*(1) = (10)^\omega$.

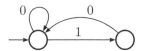

**Figure 2.19.** *An automaton for* $S_\varphi$

In Figure 2.20, the picture is slightly more complicated because $d_{\varphi^2}^*(1)$ has a different structure. The admissible paths are those that do not contain a factor belonging to the language $21^*2$.

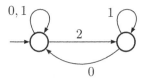

**Figure 2.20.** *An automaton for* $S_{\varphi^2}$

What we have observed in the previous example is indeed general. Below we give the general structure of the corresponding automata. For more details, see, for instance,

[LOT 02, Chapter 7]. To apply Parry's theorem, the idea is to detect a factor which is a prefix of the maximal admissible word $d_\beta^*(1)$. This corresponds exactly to the combinatorial condition about the shifted sequences.

COROLLARY 2.71.– Let $\beta > 1$ be a simple Parry number: $d_\beta(1) = t_1 \cdots t_m$ is finite, i.e. $t_m \neq 0$ and $t_j = 0$ for all $j > m$, and $d_\beta^*(1) = (t_1 \cdots t_{m-1}(t_m - 1))^\omega$. An infinite word belongs to the $\beta$-shift $S_\beta$ if and only if it is the label of a path in the automaton depicted in Figure 2.21.

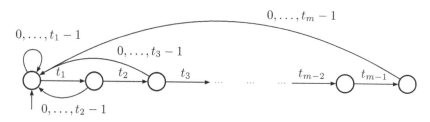

**Figure 2.21.** *An automaton for $S_\beta$ when $\beta$ is a simple Parry number*

COROLLARY 2.72.– Let $\beta > 1$ be a non-simple Parry number:

$$d_\beta(1) = d_\beta^*(1) = t_1 \cdots t_m (t_{m+1} \cdots t_{m+p})^\omega$$

where $m, p$ are taken to be minimal. An infinite word belongs to the $\beta$-shift $S_\beta$ if and only if it is the label of a path in the automaton depicted in Figure 2.22.

The following result was obtained independently by Schmidt and Bertrand [BER 77, SCH 80]. As mentioned in [BOY 96], the ideas of the proof can be traced back to [GEL 59]. A *Salem number* is an algebraic integer $\beta > 1$ whose conjugates have modulus less than or equal to 1.

THEOREM 2.73.– Let $\beta > 1$ be a Pisot number. The set of real numbers in $[0, 1]$ having an ultimately periodic $\beta$-expansion is $\mathbb{Q}(\beta) \cap [0, 1]$.

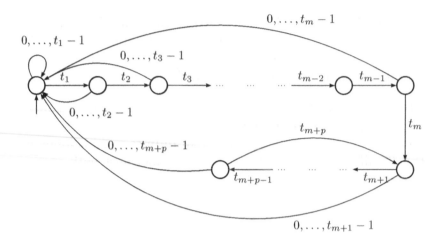

**Figure 2.22.** *An automaton for $S_\beta$ when $\beta$ is a non-simple Parry number*

Let $\beta > 1$ be a real number. If the $\beta$-expansion of every rational number in $[0, 1]$ is ultimately periodic, then $\beta$ is either a Pisot or a Salem number.

Finally, if $\beta > 1$ is an algebraic integer that is neither a Pisot nor a Salem number, then the set of rational numbers having an ultimately periodic $\beta$-expansion is nowhere dense in $[0, 1]$.

In particular, Pisot numbers are Parry numbers. Nevertheless, it is not known whether or not all Salem numbers are Parry numbers. See, for instance, [BOY 89]. However, Parry numbers are Perron numbers [DEN 76].

### 2.5.2. *Link between representations of integers and real numbers*

Bertrand's theorem stated below gives a strong link between representations of integers in a PNS and $\beta$-expansion. Recall that Bertrand numeration systems are introduced in definition 2.65.

THEOREM 2.74.– [BER 89] Let $U = (U_n)_{n \geq 0}$ be an increasing sequence of integers such that $U_0 = 1$ and $\sup_n \frac{U_{n+1}}{U_n}$ is bounded. There exists a real number $\beta > 1$ such that

$$0^* \operatorname{rep}_U(\mathbb{N}) = \operatorname{Fac}(D_\beta)$$

if and only if $U$ is a Bertrand numeration system. In that case, if $d_\beta^*(1) = t_1 t_2 \cdots$, then

$$U_n = t_1 U_{n-1} + \cdots + t_n U_0 + 1 \quad \forall n \geq 0. \tag{2.14}$$

Note that in [2.14], the $n$ht term of the sequence $U$ depends on all the previous elements. Moreover, the sequence $U$ is increasing because $t_1 \geq 1$ and all the $t_i$'s are non-negative.

Again, in Bertrand's theorem if the real number $\beta$ is a Parry number, then [2.14] has a special form. It is not difficult to see that this sequence satisfies a linear recurrence relation whenever $d_\beta^*(1)$ is ultimately periodic. Therefore, we can associate a linear numeration system built on the linear recurrent sequence $U$ with every Parry number $\beta > 1$.

If $d_\beta(1) = t_1 \cdots t_m$ is finite ($t_m \neq 0$), the sequence $(U_n)_{n \geq 0}$ defined in [2.14] satisfies

$$U_n = t_1 U_{n-1} + \cdots + t_m U_{n-m} \quad \text{for all } n \geq m,$$

$U_0 = 1$ and, for $i \in [\![1, m-1]\!]$, $U_i = t_1 U_{i-1} + \cdots + t_i U_0 + 1$.

If $d_\beta(1) = t_1 \cdots t_m (t_{m+1} \cdots t_{m+p})^\omega$ ($m$ and $p$ being chosen minimal), then the sequence $(U_n)_{n \geq 0}$ defined in [2.14] satisfies, for all $n \geq m + p$,

$$U_n = t_1 U_{n-1} + \cdots + t_{m+p} U_{n-m-p} + U_{n-p} - t_1 U_{n-p-1} - \cdots$$

$$-t_m U_{n-m-p},$$

$U_0 = 1$ and, for $i \in [\![1, m+p-1]\!]$, $U_i = t_1 U_{i-1} + \cdots + t_i U_0 + 1$.

### 2.5.3. *Ito–Sadahiro negative base systems*

Let $\beta > 1$ be a real number. As an alternative to the $\beta$-expansions discussed so far, we can look at real numbers having a representation of the form[16]

$$\sum_{i=1}^{+\infty} c_i \, (-\beta)^{-i} \, .$$

Ito and Sadahiro [ITO 09] proved that every number in the interval

$$I_\beta = \left[ -\frac{\beta}{\beta+1}, \frac{1}{\beta+1} \right)$$

has a representation as above with the digits $c_i$'s belonging to $[\![0, \lceil \beta \rceil - 1]\!]$. So we do not need any special convention to handle negative real numbers. Positive as well as negative numbers are represented with the same digit set.

We introduce the $(-\beta)$-*transformation*:

$$T_{-\beta} : I_\beta \to I_\beta, \; x \mapsto -\beta x - \left\lfloor \frac{\beta}{\beta+1} - \beta x \right\rfloor \, .$$

For $x \in I_\beta$ and $i \geq 1$, we can iteratively compute the coefficients

$$c_i = \left\lfloor \frac{\beta}{\beta+1} - \beta T_{-\beta}^{i-1}(x) \right\rfloor \, .$$

Then the word $c_1 c_2 \cdots$ is said to be the $(-\beta)$-*expansion* of $x$ and we can prove that

$$x = \sum_{i=1}^{+\infty} c_i \, (-\beta)^{-i} \, .$$

---

16 We could also consider systems with a positive base and negative digits.

The greediness of the usual $\beta$-expansions is replaced with the following condition. See [ITO 09, Prop. 3].

LEMMA 2.75.– An infinite word $\mathbf{x} = (x_i)_{i\geq 1}$ over $[\![0, \lceil \beta \rceil - 1]\!]$ is the $(-\beta)$-expansion of a real number in $I_\beta$ if and only if, for all $k \geq 0$,

$$\sum_{i=1}^{+\infty} x_{i+k} (-\beta)^{-i} \in I_\beta .$$

EXAMPLE 2.76.– Consider the golden ratio $\varphi$. We have represented in the left part of Figure 2.23 the map $T_{-\varphi}$. In this special case, the interval $I_\varphi$ is equal to $[-1/\varphi, 1/\varphi^2)$ because $\varphi^2 = \varphi + 1$.

EXAMPLE 2.77.– In the right part of Figure 2.23, we have depicted the map $T_{-3}$ where $I_3 = [-3/4, 1/4)$. Let us compute the $(-3)$-expansion of $1/5$. First, we get

$$\left(T_{-3}^i(1/5)\right)_{i\geq 0} = (1/5, -3/5, -1/5, -2/5)^\omega$$

because $T^4(1/5) = 1/5$. Then we obtain the sequence $(c_i)_{i\geq 1} = (0, 2, 1, 1)^\omega$ as $(-3)$-expansion of $1/5$. The first interval $[-3/4, -3/4 + 1/3)$, the second interval $[-3/4 + 1/3, -3/4 + 2/3)$ and the third interval $[-3/4 + 2/3, 1/4)$, correspond to the digits 2, 1 and 0, respectively.

REMARK 2.78.– If a real number $x$ does not belong to $I_\beta$, then there exists $d$ such that $x/(-\beta)^d$ belongs to $I_\beta$. Therefore, it is enough to concentrate on the expansions of the real numbers in $I_\beta$. But the situation is a bit trickier. Indeed, $x = \beta^2/(\beta+1)$ does not belong to $I_\beta$, but both $x/(-\beta)$ and $x/(-\beta)^3$ belong to $I_\beta$ and it is easy to check that the $(-\beta)$-expansions of these two numbers are not the same.

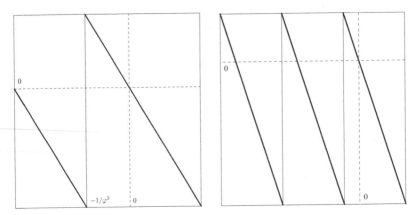

**Figure 2.23.** *The* $T_{-\varphi}$-*transformation and the* $T_{-3}$-*transformation*

There is an analogue of Parry's theorem for $(-\beta)$-expansions. First we need to define a new ordering[17].

DEFINITION 2.79.– Let $\mathbf{x} = (x_i)_{i \geq 1}$ and $\mathbf{y} = (y_i)_{i \geq 1}$ be two distinct infinite words over $\mathbb{N}$. The *alternate order* denoted by $\prec_{\mathsf{alt}}$ is defined by $\mathbf{x} \prec_{\mathsf{alt}} \mathbf{y}$ if, for the smallest $i$ such that $x_i \neq y_i$, $(-1)^i(y_i - x_i) > 0$.

For two real numbers $x, y$ in $I_\beta$, we have $x < y$ if and only if the $(-\beta)$-expansion of $x$ is less than the $(-\beta)$-expansion of $y$ for the alternate order.

By analogy with [2.13], we set

$$\mathsf{d}^*_{-\beta}\left(\frac{1}{\beta+1}\right) = \lim_{x \to (1/(\beta+1))^-} \mathsf{d}_{-\beta}(x) \,.$$

---

17 This is also the reason why we have chosen to have 1 as first index of these $(-\beta)$-expansions. We could have $(-\beta)$-expansions starting with index 0 but in that case, we must replace $(-1)^i$ with $(-1)^{i+1}$ in definition 2.79.

THEOREM 2.80.– [ITO 09] An infinite word $\mathbf{x} = (x_i)_{i \geq 1}$ over $[\![0, \lceil \beta \rceil - 1]\!]$ is the $(-\beta)$-expansion of a real number in $I_\beta$ if and only if, for all $k \geq 0$,

$$\mathsf{d}_{-\beta}\left(-\frac{\beta}{\beta+1}\right) \preceq_{\text{alt}} \sigma^k(\mathbf{x}) \prec_{\text{alt}} \mathsf{d}^*_{-\beta}\left(\frac{1}{\beta+1}\right).$$

For the base $-3$, we have

$$\mathsf{d}_{-3}(-3/4) = 3^\omega \quad \text{and} \quad \mathsf{d}^*_{-\beta}(1/4) = (02)^\omega.$$

Continuing example 2.77, let us check that we have

$$\mathsf{d}_{-3}(-3/4) = 3^\omega) \preceq_{\text{alt}} (0211)^\omega \prec_{\text{alt}} \mathsf{d}^*_{-\beta}(1/4) = (02)^\omega.$$

For the inequality on the left-hand side, for the first digit, $(-1)^1(0 - 3) \geq 0$. For the right-hand side, the first symbol where the two sequences differ is the third one and we have $(-1)^3(0 - 1) > 0$. When considering a shifted version of $(0211)^\omega$, for the inequality on the right-hand side, the two sequences $\sigma^k((0211)^\omega)$ and $(02)^\omega$ may differ with the first symbol and we have $(-1)^1(0 - 2) > 0$ (for $k = 1 + 4n$) or $(-1)^1(0 - 1) > 0$ (for $k = 2 + 4n$ or $k = 3 + 4n$).

Since the introduction of these $(-\beta)$-expansions in 2009, several related papers have appeared. First have a look at the original paper [ITO 09]. See also [MAS 11, DOM 11, LIA 12, NAK 12, AMB 12, STE 13, MAS 13].

## 2.6. Miscellaneous systems

There are many other numeration systems devised to represent integers and/or real numbers. There are even systems to represent numbers belonging to other structures (such as the complex numbers). We just give a short list with a few pointers.

Let us start with *rational base systems*. The only rational numbers that are Pisot numbers are the integers greater than one. Motivated by a number-theoretic problem, Akiyama, Frougny and Sakarovitch [AKI 08] initiated the systematic study of numeration systems associated with a rational number $p/q$. For a short presentation, the reader can also look at [BER 10, section 2.5].

Let us consider the base $3/2$. We build a PNS on the sequence $(\frac{1}{2}(\frac{3}{2})^n)_{n\geq 0}$. To decompose the non-negative integer $n_0$, we use Euclidean division (but we consider the division of $2\,n_0$ instead of $n_0$)

$$2\,n_0 = 3\,n_1 + c_0$$

with $c_0 \in \{0, 1, 2\}$. So the proposed algorithm will give the least significant digit first. Then we iterate the computation, getting

$$2\,n_i = 3\,n_{i+1} + c_i \quad \text{with } c_i \in \{0, 1, 2\}\,.$$

The procedure stops when $n_j = 0$.

Observe that the sequence $n_0, n_1, \ldots$ is decreasing; thus the procedure terminates. Therefore, we have

$$n_0 = \frac{3}{2}\,n_1 + c_0\,\frac{1}{2} = \frac{3}{2}\left(\frac{3}{2}\,n_2 + c_1\,\frac{1}{2}\right) + c_0\,\frac{1}{2} = \cdots$$

$$= \sum_{k=0}^{j-1} c_k\,\frac{1}{2}\left(\frac{3}{2}\right)^k\,.$$

As otherwise stated, every non-negative integer has such a decomposition which is the $\frac{3}{2}$-expansion of $n$.

The $\frac{3}{2}$-expansions of the first few positive integers are

2, 21, 210, 212, 2101, 2120, 2122, 21011, 21200, 21202, 21221, 210110, ... .

We can of course consider $\frac{p}{q}$-expansions of integers. Let $p, q$ be coprime integers with $1 < q < p$. Each non-negative integer can be expressed as

$$n = \sum_{k=0}^{j-1} c_k \frac{1}{q} \left(\frac{p}{q}\right)^k \quad \text{with } c_k \in [\![0, p-1]\!].$$

It is the unique finite representation of this form. Let $m, n$ be integers. We have $m < n$ if and only if the $\frac{p}{q}$-expansion of $m$ is genealogically less than that of $n$.

It is shown in [AKI 08] that the language of the $\frac{3}{2}$-expansions of the non-negative integers is prefix-closed but any two distinct subtrees in the corresponding trie are non-isomorphic. The first levels of the corresponding trie[18] are depicted in Figure 2.24. This implies that, in contrast to most situations encountered so far, this language is *not* regular (and not even context-free). In that paper, the authors also discuss the representation of real numbers.

Surprisingly, even though the language of expansions in a rational base is not regular, the computation of the successor function mapping the representation of $n$ to the one of $n+1$ can be realized by a finite transducer. The automaton depicted in Figure 2.25 reads pairs of words $(u, v)$ from right to left such that if $u$ is the $3/2$-expansions of $n$, then $v$ is the $3/2$-expansion of $n + 1$. If we do not allow leading zeroes (to have words of the same length), then the first component can be one symbol shorter. This is why there is an output function that permits an extra digit 2 in the second component.

---

18 In a breadth-first search of this trie, the labels of the corresponding edges are periodically $2, 1, 0$.

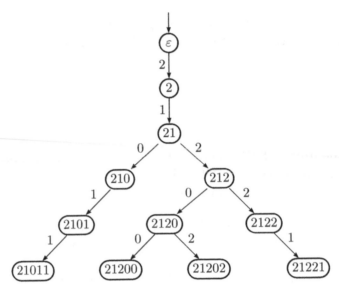

**Figure 2.24.** *A trie for words of length ≤ 5 in the prefix-closed language of the 3/2-expansions*

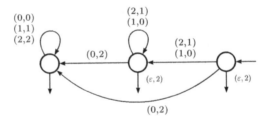

**Figure 2.25.** *Computing the successor map for 3/2-expansions*

Let $\ell \geq 2$. In the *combinatorial numeration system*, any non-negative integer can be written uniquely in the form

$$n = \binom{z_\ell}{\ell} + \binom{z_{\ell-1}}{\ell - 1} + \cdots + \binom{z_1}{1}$$

with $z_\ell > z_{\ell-1} > \cdots > z_1 \geq 0$. This system appeared in [LEH 64]. A proof of the above decomposition and its uniqueness can be obtained using ANS (the case $\ell = 2$ was indeed considered in example 2.18). See [CHA 08].

EXAMPLE 2.81.– For $\ell = 3$, we get, for instance,

$$12345678901234567890 = \binom{4199737}{3} + \binom{3803913}{2} + \binom{1580642}{1}.$$

The *factorial numeration system* is a positional system based on the sequence $(n!)_{n \geq 1}$. Every non-negative integer can be written in a unique way as

$$n = c_k\, k! + c_{k-1}\, (k-1)! + \cdots + c_2\, 2! + c_1\, 1!$$

with $c_k \neq 0$ and $c_i \in [\![0, i]\!]$ for all $i \leq k$. See, for instance, [LEN 05] where this system serves as an introduction to introduce profinite Fibonacci numbers. This factorial system is beyond the scope of this book because the digits are not bounded and we are usually looking at systems giving representations that are words over some finite alphabet.

The *Ostrowski numeration system* [OST 22] is based on the denominators of the convergents of the continued fraction expansion[19] of a real number. Let $\alpha$ be an irrational number in $(0, 1)$ with a continued fraction expansion $[0; a_1, a_2, \ldots]$. For all $n$, consider the $n$th convergent which is the rational number

$$\frac{p_n}{q_n} = \cfrac{1}{a_1 + \cfrac{1}{a_2 + \cfrac{\ddots}{{} + \frac{1}{a_n}}}}.$$

We have $q_0 = 1$, $q_1 = a_1$ and $q_{n+1} = a_{n-1}q_n + q_{n-1}$. With $\alpha$ equal to the golden ratio, we obtain the usual Zeckendorf numeration system.

---

19 See section 1.3.5, Volume 1, about continued fraction expansions.

EXAMPLE 2.82.– Take $\alpha = \sqrt{7} - 2 \simeq 0.646$. The first few convergents are

$$1,\ 1/2,\ 2/3,\ 9/14,\ 11/17,\ 20/31,\ 31/48,\ 144/223,\ 175/271,\ \ldots.$$

Hence, we consider the sequence $(q_n)_{n \geq 0} = 1, 2, 3, 14, 17, 31, 48, 223, 271, \ldots$.

PROPOSITION 2.83.– Every non-negative integer can be written in a unique way as

$$n = \sum_{k=1}^{m} d_k\, q_{k-1}$$

where $d_1 \in [\![0, a_1 - 1]\!]$, for all $k \geq 2$, $d_k \in [\![0, a_k]\!]$ and $d_k = 0$ whenever $d_{k+1} = a_{k+1}$.

For a survey about this system and its combinatorial properties, see [BER 01]. Note that we can also represent real numbers in a similar way but using the sequence $(q_n \alpha - p_n)_{n \geq 0}$. There are also some good connections with combinatorial game theory. See, for instance, [FRA 82].

To represent real numbers in $[0, 1)$, we can also consider the so-called *Cantor series expansions*. Let $Q = (q_n)_{n \geq 1}$ be a sequence of integers in $\mathbb{N}_{\geq 2}$. Every real $x \in [0, 1)$ can be written as

$$x = \sum_{n=1}^{+\infty} \frac{c_n}{q_1 \cdots q_n}$$

where $c_n \in [\![0, q_n - 1]\!]$, for all $n$, and the expansion does not eventually end with $E_n = q_n - 1$. Note that if the sequence $Q$ is constant, i.e. $q_n = b$ for all $n \geq 1$, then we are exactly in the case of the base-$b$ expansions of real numbers. We can conveniently introduce the sequence of maps

$$T_{Q,n} : [0, 1) \to [0, 1),\ x \mapsto \{q_1 \cdots q_n x\}\,.$$

See, for instance, [MAN 12] to obtain a few pointers to the existing literature.

We conclude this section with the introduction of the *shift radix systems* (SRS) which are again examples of dynamical systems. For a survey of these systems and discussion of the links existing with numeration systems, see [KIR 13].

DEFINITION 2.84.– Let $d \geq 1$ be an integer and $r = (r_0, \ldots, r_{d-1}) \in \mathbb{R}^d$. Consider the map

$$T_r : \mathbb{Z}^d \to \mathbb{Z}^d, \ z = (z_0, \ldots, z_{d-1})^\mathsf{T} \mapsto (z_1, \ldots, z_{d-1}, \lfloor \langle r, z \rangle \rfloor)^\mathsf{T}.$$

It is not difficult to see that if $z = (z_0, \ldots, z_{d-1})^\mathsf{T}$, then $T_r(z) = z' = (z_1, \ldots, z_d)^\mathsf{T}$ where $z_d$ is the unique integer satisfying $0 \leq \langle r, z \rangle - z_d < 1$.

For instance, $\beta$-expansions are conjugated to a SRS $T_r$ for a parameter $r$ depending on $\beta$ and Rauzy fractals can also be obtained as a geometric representation of some SRS. In the theory of (topological) dynamical systems, two dynamical systems $(X_1, f_1)$ and $(X_2, f_2)$ are *conjugate* if there exists a homeomorphism $g : X_1 \to X_2$ such that $g \circ f_1 = f_2 \circ g$.

## 2.7. Bibliographical notes and comments

There are several **surveys** developing other directions about numeration systems. See [BAR 06a] for a unifying approach and several applications of numeration systems (in particular, discussion of the so-called odometer). Dynamical properties (ergodicity, chaoticity) for computer arithmetic purposes are surveyed in [BER 12]. In particular, the amount of information contained in some truncated expansion is also discussed. My survey [RIG 10] presents topics similar to those presented in this book (with an extension to $b$-recognizable sets of real numbers by Büchi automata and some applications to number theory). The survey [BRU 94] is oriented more to the main subject of Chapter 3: application of first-order logic to $b$-recognizable sets of integers. A gentle presentation of several systems of numeration can also be found in [FRA 85]. Another survey about abstract numeration systems can be found in [RAM 11], see also [BER 10, Chapter 3].

There are several **variants of integer base systems**. Let $b \geq 2$ be an integer. In the so-called *Avizienis numeration system* [AVI 61], is a redundant system where the digit-set $[\![0, b-1]\!]$ is replaced with $[\![-a, a]\!]$ where $a < b$ and $2a \geq b+1$. For practical reasons, such a system can be used to compute addition in parallel in constant time (there is no carry propagation). See also [FRO 11]. The survey paper [FRO 07] is also of interest with respect to computer arithmetic. In particular, Frougny observed that Cauchy was already using the digit-set $\{-5, \ldots, 5\}$ in base 10. In the same survey paper, we can find a description of the Booth transducer used to replace usual base-2 expansions with Avizienis expansions over the digit-set $\{-1, 0, 1\}$: every factor $01^n$ is changed into $10^{n-1}(-1)$.

Of course, [KNU 12, Chapter 4] about number theory and positional numeration systems is also worth reading. It also contains a few pages about historical perspectives. The complex number $\beta = -1 + i$ can be used as a base to decompose every Gaussian integer $z$ in $\mathbb{Z}[i]$ as $z = c_h \beta^h + \cdots + c_0$ with $c_i \in \{0, 1\}$ and $c_h = 1$ in a unique way. See [KNU 60] and [PEN 65]. This has led to many generalizations. See [GIL 81] and then [KÁT 81] and [PET 91] where the notion of **canonical number system** was introduced. Let $(\beta, D)$ be a pair where $\beta \in \mathbb{C}$ is an algebraic integer of modulus greater than 1 and $D \subset \mathbb{Z}[\beta]$ is a finite alphabet containing 0 and playing the role of digit-set.

This pair is a canonical number system if every element in $\mathbb{Z}[\beta]$ has a unique decomposition

$$\sum_{k=0}^{\ell} d_k \beta^k$$

with $d_\ell \neq 0$ and $d_k \in D$ for all $k$.

Theorem 2.24 (Cobham's theorem on automatic sequences) for **abstract numeration systems** was extended to a multi-dimensional setting in [CHA 10].

About various enumeration of points in the plane such as the **Peano enumeration**, automaticity of the resulting sequence can be considered. Let $(a_{i,j})_{i,j\geq 0}$ be a two-dimensional array. If $(a_{i,j})_{i,j\geq 0}$ is $k$-automatic, then its "linearization" $1 = a_{00}a_{01}a_{10}a_{02}a_{11}a_{20}a_{03}\cdots$ need not necessarily be $k$-automatic [ALL 03a, p. 424]. Also see the last section of [NIC 07].

Let $U = (U_n)_{n\geq 0}$ be a positional numeration system. We have seen that if $\mathrm{rep}_U(\mathbb{N})$ is regular, then $U$ satisfies a linear recurrence relation. Let $U$ be a **linear numeration system**. Sufficient conditions for $\mathbb{N}$ to be $U$-recognizable are considered in [LOR 95] and [HOL 98].

There is also plenty to say about the so-called **odometer** and **successor function**. Given a numeration system (e.g. a positional numeration system $U$ or an abstract numeration system $S$), the *successor function* maps the representation of the integer $n$ to the representation of $n+1$. So we are dealing with a map defined over a set of finite words. In [FRO 97] a characterization of the sequences $U = (U_n)_{n\geq 0}$ such that the successor function is a left (respectively, right) sequential function, i.e. computed by a finite transducer reading from most (respectively, least) significant first, is provided. Considering the extension of finite representations to infinite words, we can define the odometer which is the natural extension of the successor map in such a setting. See, for instance, [GRA 95] and [BAR 06a, BER 07d]. As we demonstrate, consider the set of infinite words over $[\![0, 9]\!]$ and the function that maps $9^n j$w to $9^n (j + 1)$w, for all $n \geq 0$ and $j < 9$, and that maps $9^\omega$ to $0^\omega$. As another example, consider the set of infinite words over $\{0, 1\}$ avoiding the factor 11, we can define the odometer acting on this set as $(01)^n 00$w $\mapsto$ $(00)^n 10$w, for all words w avoiding 11, and $(01)^\omega \mapsto 0^\omega$. The reader can think about studying carry propagation in the Zeckendorf system (where representations are written with least significant digit first) extended to infinite words.

Theorem 2.60 was considering the normalization of $U$-representations of integers. **Normalization** of $\beta$-expansions (of real numbers) can be computed by a finite automaton if and only if $\beta$ is a Pisot number. See [BER 94a].

Many papers are devoted to $\beta$-expansions. More results and several pointers on $\beta$-expansions can be found in [LOT 02, section 7.2]. In particular, $\beta$-**integers** were extensively studied [THU 89]. See, for instance, [BER 08, BER 07a, MAS 04, FRO 03, BUR 98]. We have mostly considered $\beta$-expansions of real numbers in $[0, 1)$. Nevertheless, multiplying by a convenient power of $\beta$, we can represent every positive real number as follows. Let $x \geq 1$. There exists a smallest $n$ such that $\beta^{-n} x < 1$. If $d_\beta(\beta^{-n} x) = c_1 c_2 \cdots$, then the $\beta$-expansion of $x$ is $c_1 \cdots c_n \bullet c_{n+1} \cdots$.

A non-negative (respectively, negative) real number $x$ is a $\beta$-*integer* if its $\beta$-expansion (respectively, the $\beta$-expansion of $-x$) is of the form $c_1 \cdots c_n \bullet 0^\omega$. (So to speak, $x$ is a $\beta$-integer if its $\beta$-fractional part is zero.) Equivalently, the non-negative $\beta$-integers are the real numbers $x = \sum_{k=0}^n c_i \beta^i$ with $c_i \in [\![0, \lceil \beta \rceil - 1]\!]$ such that $c_j \cdots c_0$ is genealogically less then or equal to the prefix of $d_\beta^*(1)$ of length $j + 1$ for all $j \in [\![0, k]\!]$. In Figure 2.26, we have represented the first few $\varphi$-integers where $\varphi$ is the golden ratio:

$$1, \varphi, \varphi^2, \varphi^2 + 1, \varphi^3, \varphi^3 + 1, \varphi^3 + \varphi, \ldots .$$

We can notice that the distances between two consecutive $\varphi$-integers take only two values. If these two values are coded by a and b, we get the Fibonacci word again.

● ●● ● ●● ●● ● ●● ● ●● ●● ● ●● ●●

**Figure 2.26.** *The first few positive $\varphi$-integers*

In a similar way, we can be interested in numbers with finite $\beta$-expansions. See, for instance, [FRO 92b].

# 3

# Logical Framework and Decidability Issues

The study of the connections between mathematical
automata and formal logic is as old as theoretical
computer science itself. [...] Research on the logical
aspects of the theory of finite-state automata began in
the early 1960s with the work of J. Richard Büchi on
monadic second-order logic.

Howard Straubing [STR 94]

The aim of this chapter is to explain that every
$k$-automatic word (or, equivalently, every $k$-recognizable set of
non-negative integers) can be defined using a first-order
formula in an appropriate logical structure that is an
extension of the Presburger arithmetic (of course, we have to
make such a statement explicit). It turns out that this
important result due to Büchi has interesting applications in
proving theorems about $k$-automatic words using an
algorithmic procedure. Such a technique can be generalized
to $U$-recognizable sets of integers within a Pisot numeration
system as described in section 2.4.

Even if time or space complexity issues may appear (but
these problems will not be discussed much here), the term

*automatic theorem-proving*[1] means that after writing an appropriate formula, we simply have to carry out some algorithm that permits us to decide in a finite amount of time whether or not the expressed property holds. Of course, such a result also has its own limitations (the automata that are produced can be quite large and not every property about $k$-automatic words can be expressed using this formalism). As an example, we can obtain an automatic certification that the overlap-freeness of the Thue–Morse word holds. In the same spirit, the following problem turns out to be decidable: given any uniform morphism and any coding generating a $k$-automatic word, decide whether or not this word is ultimately periodic.

Let us give a glimpse of the arguments we will be dealing with in this chapter. The existence of an overlap occurring in an infinite word w corresponds to the satisfiability of the following formula, where the variables are non-negative integers indexing positions within $\mathbf{w} = \mathbf{w}(0)\mathbf{w}(1)\mathbf{w}(2)\cdots$

$$(\exists i \geq 0)(\exists \ell \geq 1)(\forall j \in [\![0, \ell]\!])(\mathbf{w}(i + j) = \mathbf{w}(i + \ell + j)).$$

Indeed, we express with $j \in [\![0, \ell - 1]\!]$ the fact that a square of length $2\ell$ is detected in position $i$, i.e. $w_i \cdots w_{i+\ell-1} = w_{i+\ell} \cdots w_{i+2\ell-1}$ and with $j = \ell$ we express the occurrence of an overlap (the first letter has to be repeated three times), i.e. $w_{i+\ell} = w_{i+2\ell}$.

The readers are not assumed to have any particular background in mathematical logic. All the prerequisites about mathematical logic are rigorously presented in section 3.1. Assuming that a naive notion of algorithm is at hand (we will not introduce Turing machines), decision problems and decidable problems are defined in section 3.2.

---

1 In reference to [GOČ 12].

In the third section, we present Presburger arithmetic: in this first-order theory, we have only access to addition of (non-negative) integer variables, test for equality and quantification of variables ranging over $\mathbb{N}$. We discuss the decidability of this theory using the classical technique of the elimination of quantifiers. The idea is that from every sentence, i.e. a formula where all variables are under the scope of a quantifier, we can get an equivalent sentence without any quantifier, and thus we have "simply" to check some equality relations about integers. In particular, we study the structure of the subsets of $\mathbb{N}^n$ that can be described by a formula in $\langle \mathbb{N}, + \rangle$.

Section 3.4 is the heart of this chapter. We present Büchi's theorem: a set $X \subseteq \mathbb{N}^n$ is $k$-recognizable if and only if it is definable in $\langle \mathbb{N}, +, V_k \rangle$ where $V_k(n)$ is the largest power of $k$ dividing $n$. The first-order theory of this latter structure is again decidable. The key arguments are that all the atomic relations are representable by finite automata and languages recognized by finite automata are closed under Boolean operations (see theorems 1.8 and 1.9). Then, we discuss some corollaries of Büchi's theorem and possible applications to combinatorics on words such as proving or disproving the occurrence of repetitions or overlaps.

> Warning: I do not want the readers to be discouraged. Section 3.1 and section 3.2 serve as reference sections with all the necessary formal presentation leading to rigorous constructions. If the readers just want to get a feeling about the connections between $k$-automatic words and logic, these two sections can be skipped on a first reading. We can quickly browse through section 3.3 and then go to page 156.

## 3.1. A glimpse at mathematical logic

Our first aim is to build syntactically correct formulas and then we will interpret them. Do not be afraid: at first sight, things are a bit formal. But in the next few pages, I hope that the interpretation of the structures that we will constantly use should be clear.

Our purpose is two-fold. First, we will build syntactically correct formulas, so we will give rules to inductively produce well-formed "words". These words will make a (first-order) language. On the other hand, we will interpret these formulas and be concerned with semantics: we will add some meaning. This section is inspired by the presentation given in [EBB 94], where the readers can find much more material about mathematical logic and an introduction to model theory.

### 3.1.1. *Syntax*

A *structure* $\mathfrak{s} = \langle D, (f_i)_{i \in I}, (R_j)_{j \in J} \rangle$ is given by a set $D$, a family $(f_i)_{i \in I}$ of symbols for functions and a family $(R_j)_{j \in J}$ of symbols for relations that are defined over $D$. The set $D$ is called the *domain* (or the underlying set) of the structure.

With each functional symbol $f_i$ (respectively, relational symbol $R_j$) is associated an *arity* that is a non-negative integer (respectively, a positive integer). A function $f_i$ of arity $0$ is a constant belonging to $D$. A function of arity $n \geq 1$ is a map whose domain is $D^n$. A relation of arity $n \geq 1$ is a subset of $D^n$.

REMARK 3.1.– At this early stage, we will make a distinction between the symbol $f_i$ denoting a function (respectively, a relation $R_j$) of arity $n$ and the function $f_i^{\mathfrak{s}} : D^n \to D$ (respectively, the relation $R_j^{\mathfrak{s}} \subseteq D^n$) itself that is the *interpretation* of the symbol within the structure $\mathfrak{s}$. In this context, the set of symbols denoting the functions and relations is called a *symbol set*.

Otherwise stated, a structure is given by a domain $D$, a symbol set $S$ and an interpretation of the elements in the symbol set. Usually, the interpretation is clear from the context.

EXAMPLE 3.2.– Take the symbol set $S = \{+, \times, 0, =, <\}$ with five symbols where it is assumed that we have two functions $+$ and $\times$ of arity 2, a function 0 of arity 0 and two relations $=$ and $<$ of arity 2. For the structure $\mathfrak{q} = \langle \mathbb{Q}, +, \times, 0, =, < \rangle$ a natural interpretation is to consider $+^{\mathfrak{q}} : \mathbb{Q} \times \mathbb{Q} \to \mathbb{Q}$ (respectively, $\times^{\mathfrak{q}} : \mathbb{Q} \times \mathbb{Q} \to \mathbb{Q}$) as the usual addition (respectively, multiplication) of rational numbers, the interpretation $0^{\mathfrak{q}}$ is the number $0 \in \mathbb{Q}$ and $=^{\mathfrak{q}}$ and $<^{\mathfrak{q}}$ are the usual equality and order relations over $\mathbb{Q}$.

Take the same symbol set $S$, but consider the structure $\mathfrak{n} = \langle \mathbb{N}, +, \times, 0, =, < \rangle$ with the natural interpretation of the operations and relations over $\mathbb{N}$. For instance, $+^{\mathfrak{n}} : \mathbb{N} \times \mathbb{N} \to \mathbb{N}$ is the usual addition of non-negative integers.

If the interpretation is clear from the context (and usually, we will not change the interpretation), we will not distinguish a symbol from its interpretation within the structure.

Let $S$ be a symbol set with two families $(f_i)_{i \in I}$ and $(R_j)_{j \in J}$ of symbols for functions and relations (with their own arity). We assume moreover that $=$ is always a binary relation belonging to $S$ with its usual interpretation as equality.

We extend $S$ to define the *alphabet of a first-order language $A_S$*. We add countably many variables $x_1, x_2, x_3, \ldots$ (or $x, y, z, \ldots$) and extra symbols $\vee, \wedge, \to, \leftrightarrow, \neg, \exists, \forall$ not in $S$ as well as parentheses. We define terms and formulas inductively, and we often make use of extra parentheses for the sake of clarity of the constructions.

DEFINITION 3.3 (Terms).– First, we define *terms* that are particular words over $A_S^*$. To construct them, we may apply the following rules finitely many times.

– Any variable is a term.

– Any constant in $S$, i.e. any function symbol of arity $0$, is a term.

– If $t_1, \ldots, t_n$ are terms and if $f_i \in S$ is a function symbol of arity $n$, then $f_i(t_1, \ldots, t_n)$ is a term.

EXAMPLE 3.4.– Assume that $S$ contains two symbols $f_1, f_2$ for some binary functions and a symbol $c$ for a constant. We list a few examples of terms over $S$

$$x_1, \quad c, \quad f_1(x_1, x_2), \quad f_2(c, c), \quad f_1(x_1, f_1(f_2(c, x_1), x_2)).$$

DEFINITION 3.5 (Formulas).– Next, we define *formulas* inductively by applying the following rules finitely many times. Note that these formulas are also words over $A_S^*$.

– If $t_1, \ldots, t_n$ are terms and if $R_j \in S$ is a relation symbol of arity $n$, then $R_j(t_1, \ldots, t_n)$ is a formula. In particular, if $s$ and $t$ are terms, then $s = t$ is a formula.

– If $\varphi$ and $\psi$ are formulas, then $\varphi \vee \psi$, $\varphi \wedge \psi$, $\varphi \to \psi$, $\varphi \leftrightarrow \psi$, $\neg\varphi$ are formulas.

– If $\varphi$ is a formula and $x$ is a variable, then $\forall x \varphi$ and $\exists x \varphi$ are formulas.

If a formula is obtained by applying only the first rule, then it is called an *atomic formula*. The set of formulas over $A_S$ is the *first-order[2] language associated with $S$*.

EXAMPLE 3.6.– Let us continue with example 3.4. The symbol set $S$ contains $f_1, f_2, c$ but also a relation symbol $R$ for

---

2 The terminology "first-order" means that universal or existential quantifiers are only applied to variables. It is not allowed to quantify over sets of variables or more generally over relations.

a ternary relation, i.e. with arity $3$. We list a few examples of atomic formulas

$$R(x_1, x_2, x_3), \quad R(c, x_1, f_1(x_1, x_2)), \quad x_1 = f_2(x_2, x_3).$$

For the third one, recall that we always assume that equality is a binary relation belonging to $S$. For usual relations like equality, we write $x_1 = f_2(x_2, x_3)$ instead of $= (x_1, f_2(x_2, x_3))$. Here, are some formulas

$$R(x_1, x_2, x_3) \vee R(c, x_1, f_1(x_1, x_2)), \quad \exists x_2 f_2(x_2, x_3)$$

and

$$\forall x_1 \forall x_2 \exists x_3 ((R(x_1, x_2, x_3) \wedge R(c, x_1, f_1(x_1, x_2)))$$
$$\rightarrow R(x_1, x_1, c)). \qquad\qquad [3.1]$$

For this last example, we have added extra parentheses to increase the readability of the whole formula (and to avoid any ambiguity).

A *sentence* (sometimes called a *closed formula*) is a formula where each variable is within the scope of some quantifier. As an example, $(\forall x)(\exists y)(x = y + y)$ is a sentence because the variables $x$ and $y$ are in the scope of a universal and existential quantifier, respectively. But the formula $(\exists z)(x + z = y)$ is not a sentence because the variables $x$ and $y$ are *free*: they are not in the scope of any quantifier. Sometimes, it is convenient to highlight the presence of free variables, so we can write $\varphi(x, y) \equiv (\exists z)(x + z = y)$. We use the notation $\equiv$ to provide an explicit description of a formula and to avoid any confusion with the equality symbol that is often used inside formulas.

Free variables of a formula can be defined inductively. We denote by $\text{Free}(\varphi)$ the set of free variables occurring in $\varphi$. Let $S$ be a symbol set.

– If $t_1, \ldots, t_n$ are terms and if $R_j \in S$ is a relation symbol of arity $n$, then $\mathsf{Free}(R_j(t_1, \ldots, t_n))$ is the union of the sets of variables occurring in the different terms $t_i$. In particular, if $s$ and $t$ are terms, then $\mathsf{Free}(s = t)$ is made up of the variables of $s$ and $t$.

– If $\varphi$ and $\psi$ are formulas, then

$$\mathsf{Free}(\varphi \vee \psi) = \mathsf{Free}(\varphi \wedge \psi) = \mathsf{Free}(\varphi \to \psi) = \mathsf{Free}(\varphi \leftrightarrow \psi)$$
$$= \mathsf{Free}(\varphi) \cup \mathsf{Free}(\psi)$$

and $\mathsf{Free}(\neg\varphi) = \mathsf{Free}(\varphi)$.

– If $\varphi$ is a formula and $x$ is a variable, then

$$\mathsf{Free}(\forall x\varphi) = \mathsf{Free}(\exists x\varphi) = \mathsf{Free}(\varphi) \setminus \{x\}\,.$$

EXAMPLE 3.7.– Consider the formula where, for each subformula, we have written the corresponding set of free variables,

$$\exists x_3(( \underbrace{R(x_1, x_2, x_3)}_{\mathsf{Free}(\cdot)=\{x_1,x_2,x_3\}} \wedge \underbrace{(\forall x_2 \overbrace{R(c, x_1, f_1(x_1, x_2))}^{\mathsf{Free}(\cdot)=\{x_1,x_2\}})}_{\mathsf{Free}(\cdot)=\{x_1\}}) \to \underbrace{(\forall x_1 \overbrace{R(x_1, x_1, c)}^{\mathsf{Free}(\cdot)=\{x_1\}})}_{\mathsf{Free}(\cdot)=\emptyset})\,.$$

$$\underbrace{\phantom{\exists x_3(( R(x_1, x_2, x_3) \wedge (\forall x_2 R(c, x_1, f_1(x_1, x_2))) \to (\forall x_1 R(x_1, x_1, c)))}}_{\mathsf{Free}(\cdot)=\{x_1,x_2,x_3\}}$$

Its free variables are $x_1$ and $x_2$. In the formula [3.1], there is no free variable hence, [3.1] is a sentence.

## 3.1.2. *Semantics*

Let $\mathfrak{s}$ be a structure with domain $D$ and symbol set $S$. Up to now, we have described what are the syntactically correct formulas in the first-order language associated with $S$. Now, we turn to their interpretation and define the meaning of $\mathfrak{s} \models \varphi$ where $\varphi$ is a formula[3]. Let us start with an example.

---

3 The notation $\mathfrak{s} \models \varphi$ can be read "$\varphi$ is satisfied in $\mathfrak{s}$".

EXAMPLE 3.8.– Consider the structures introduced in example 3.2. For the structure $q = \langle \mathbb{Q}, +, \times, 0, =, < \rangle$, the sentence[4]

$$\varphi \equiv \forall x \exists y (x = y + y)$$

should be interpreted as follows. For every rational number $x \in \mathbb{Q}$, there exists a rational number $y \in \mathbb{Q}$ such that $x = 2y$. Clearly, this interpretation holds: for all $x \in \mathbb{Q}$, we take $y = x/2$. We will write $q \models \varphi$.

Now consider the structure $n = \langle \mathbb{N}, +, \times, 0, =, < \rangle$ and the same sentence $\varphi$. In this case, the interpretation of $\varphi$ is the statement "For every number $x \in \mathbb{N}$, there exists a number $y \in \mathbb{N}$ such that $x = 2y$" that does not hold (indeed, if $x$ is odd, there is no such integer $y$). We will write $n \not\models \varphi$.

Finally, consider the formula

$$\psi(x) \equiv \exists y (x = y + y)$$

with a free variable $x$. Recall that the notation $\psi(x)$ reflects the fact that $x$ is a free variable occurring in $\psi$. If we replace $x$ with an even integer $2n$, i.e. we substitute $x$ with a specific even value, then the formula $\psi(2n)$ is satisfied when interpreted in $n$. We write $(n, x \mapsto 2n) \models \psi(x)$. On the other hand, if we replace $x$ with an odd integer $2n+1$, then the formula $\psi(2n+1)$ is not satisfied when interpreted in $n$. We write $(n, x \mapsto 2n + 1) \not\models \psi(x)$. Otherwise stated, the set

$$\{n \in \mathbb{N} \mid (n, x \mapsto n) \models \psi(x)\}$$

is the set of even integers. As a conclusion, observe that the validity of the interpretation, when free variables occur, may depend on the values assigned to these free variables.

---

4 Recall that the $\equiv$ is not in the symbol set. We use this symbol instead of $=$ to define a formula.

Let $\mathfrak{s} = \langle D, (f_i)_{i \in I}, (R_j)_{j \in J} \rangle$ be a structure having $S = \{(f_i)_{i \in I}, (R_j)_{j \in J}\}$ as symbol set.

We are now concerned with interpretation. Every constant $c$ in $S$ is interpreted by an element $c^{\mathfrak{s}} \in D$. In the same way, every function $f_i$ has its interpretation $f_i^{\mathfrak{s}}$. So, every term where no variable occurs can be interpreted as an element in $D$. If some variable occurs, we need to define an *assignment*, i.e. a map $\alpha$ from the set $\{x_1, x_2, x_3, \ldots\}$ of variables to $D$. If such an assignment $\alpha$ is given, then every variable is substituted with an element in $D$ and every term $t$ can be interpreted as an element in $D$ denoted by $t^{(\mathfrak{s}, \alpha)}$.

Let $\varphi, \psi$ be two formulas in the first-order language associated with the structure $\mathfrak{s}$. Let $\alpha$ be an assignment. Since formulas were defined inductively, we follow exactly the same scheme to define the meaning of $(\mathfrak{s}, \alpha) \models \varphi$.

– Assume that $\varphi \equiv R_j(t_1, \ldots, t_n)$ where $t_1, \ldots, t_n$ are terms and $R_j \in S$ is a relation of arity $n$, i.e. $R_j$ is a subset of $D^n$. Then $(\mathfrak{s}, \alpha) \models \varphi$ if and only if $((t_1)^{(\mathfrak{s}, \alpha)}, \ldots, (t_n)^{(\mathfrak{s}, \alpha)}) \in R_j^{\mathfrak{s}}$. In particular if $s, t$ are terms, then $(\mathfrak{s}, \alpha) \models s = t$ if and only if $s^{(\mathfrak{s}, \alpha)} = t^{(\mathfrak{s}, \alpha)}$.

– $(\mathfrak{s}, \alpha) \models \varphi \vee \psi$ if and only if $(\mathfrak{s}, \alpha) \models \varphi$ or $(\mathfrak{s}, \alpha) \models \psi$;

– $(\mathfrak{s}, \alpha) \models \varphi \wedge \psi$ if and only if $(\mathfrak{s}, \alpha) \models \varphi$ and $(\mathfrak{s}, \alpha) \models \psi$;

– $(\mathfrak{s}, \alpha) \models \neg\varphi$ if and only if $(\mathfrak{s}, \alpha) \not\models \varphi$, i.e. $(\mathfrak{s}, \alpha) \models \varphi$ does not hold;

– $(\mathfrak{s}, \alpha) \models \varphi \rightarrow \psi$ if and only if $(\mathfrak{s}, \alpha) \models \neg\varphi \vee \psi$;

– $(\mathfrak{s}, \alpha) \models \varphi \leftrightarrow \psi$ if and only if $(\mathfrak{s}, \alpha) \models \varphi \rightarrow \psi$ and $(\mathfrak{s}, \alpha) \models \psi \rightarrow \varphi$;

– $(\mathfrak{s}, \alpha) \models \forall x \varphi$ if and only if, for all $d \in D$ and all assignments

$$\alpha_d : \begin{cases} x \mapsto d, \\ y \mapsto \alpha(y), & \text{if } y \neq x \end{cases} \qquad [3.2]$$

we have $(\mathfrak{s}, \alpha_d) \models \varphi$;

$- (\mathfrak{s}, \alpha) \models \exists x \varphi$ if and only if there exists $d \in D$ and the corresponding assignment $\alpha_d$ defined as in [3.2] is such that $(\mathfrak{s}, \alpha_d) \models \varphi$.

Let $n \in \mathbb{N}_{\geq 1}$ and $d_1, \ldots, d_n \in D$. Let $\varphi = \varphi(x_1, \ldots, x_n)$ be a formula with $n$ free variables $x_1, \ldots, x_n$. We write $\mathfrak{s} \models \varphi(d_1, \ldots, d_n)$ if and only if $(\mathfrak{s}, \beta) \models \varphi$ where $\beta$ is any assignment such that $\beta(x_i) = d_i$ for all $i = 1, \ldots, n$. We can, therefore, define a subset of $D^n$

$$M_{\mathfrak{s}, \varphi} := \{ (d_1, \ldots, d_n) \in D^n \mid \mathfrak{s} \models \varphi(d_1, \ldots, d_n) \}. \qquad [3.3]$$

If $M_{\mathfrak{s}, \varphi}$ is non-empty, then $\varphi$ is said to be *satisfiable*. Otherwise stated, $M_{\mathfrak{s}, \varphi}$ is the set of $n$-tuples satisfying $\varphi$.

EXAMPLE 3.9.– Again consider the structure $\mathfrak{n} = \langle \mathbb{N}, +, \times, 0, =, < \rangle$ and the formula $\psi(x) \equiv \exists y (x = y + y)$. With the above notation, we have

$$M_{\mathfrak{n}, \psi} = \{ m \in \mathbb{N} \mid \mathfrak{n} \models \psi(m) \} = 2\mathbb{N}.$$

Thus, the formula $\psi$ is satisfiable (within $\mathfrak{n}$).

DEFINITION 3.10.– Let $\mathfrak{s}$ and $\mathfrak{t}$ be two structures with the same domain. These structures are *equivalent* if, for all formulas $\varphi$ of $\mathfrak{s}$, there exists a formula $\tau$ of $\mathfrak{t}$ such that $M_{\mathfrak{s}, \varphi} = M_{\mathfrak{t}, \tau}$ and conversely. Two formulas $\varphi$ and $\tau$ in $\mathfrak{s}$ are *equivalent* if $M_{\mathfrak{s}, \varphi} = M_{\mathfrak{s}, \tau}$.

DEFINITION 3.11.– Let $\mathfrak{s}$ be a structure with domain $D$ and symbol set $S$. Let $L_0(S)$ be the set of sentences over $A_S$. The (first-order) *theory* of $\mathfrak{s}$ is the set

$$\{ \varphi \in L_0(S) \mid \mathfrak{s} \models \varphi \}.$$

Instead of relation, we can also use the term *predicate*. Usually, a predicate is assumed to be the characteristic function of a relation.

## 3.2. Decision problems and decidability

We give a quick overview of decision problems and decidability. There are several textbooks about computability and computational complexity. See, for instance, [ARO 09, SUD 06, PAP 94].

Roughly speaking, when asking a question involving one or several parameters taking infinitely[5] many values and where the expected answer can only be yes/no, then we are dealing with a decision problem. An *instance* of such a problem is a given input, i.e. particular values are given to the parameters.

DEFINITION 3.12.– A *decision problem* $P$ can be formalized as a map from an infinite set $I$, i.e. the collection of instances, or inputs, to the binary set $\{yes, no\}$. By choosing a convenient coding[6], elements of $I$ are replaced by words over some finite alphabet $A$. The codings of the positive instances of the problem define some formal language $L_P$. It is quite usual to make no distinction between the problem $P$ and the language $L_P$. Also, it is usual to consider the partition of $A^*$ into $L_P$ and $A^* \setminus L_P$ where this latter set contains the codings of negative instances as well as the words in $A^*$ that are not a valid coding.

DEFINITION 3.13.– A problem $P$ with a set $I$ of instances is *decidable* if there exists an algorithm (we will not precisely define here what an algorithm is, i.e. a Turing machine) that

---

5 If the set of instances is finite, then there is always a finite algorithm (corresponding to the finite list of positive instances) that solves the problem.

6 We can always find a convenient description: integers can be represented as words over a finite alphabet of digit within some numeration system, first-order formulas are words over some alphabet $A_S$, a deterministic finite automation (DFA) can be described by its transition function, its initial state and the set of final states; a morphism $f : A^* \to B^*$ can also be specified by a finite word and some conventions, e.g. a string like #a#ab#b#ba# can be used to describe the Thue–Morse morphism, . . .

for every input $x \in I$ coded by a word $x' \in A^*$ will halt in a finite number of steps and tell whether or not $x'$ is the coding of a positive instance, i.e. $x' \in L_P$.

For the readers used to the formalism of Turing machines, this means that the language $L_P$ is recursive. Roughly speaking, there exists an algorithm that for every input string $w$ decides whether or not $w$ belongs to $L_P$. Since we make no distinction between a decision problem $P$ and the language $L_P \subseteq A^*$, we can also speak of a *decidable language* $L$: there exists an algorithm that always halts on every input in $A^*$ and accepts exactly the words in $L_P$.

EXAMPLE 3.14 (Some decidable problems).– As an example, consider the question PRIMES: is $n \in \mathbb{N}$ a prime number? Take, for instance, the base-2 expansions of the positive integers as a coding. If we make the distinction between the positive instances from the negative ones, we get the following language

$$L_{\text{PRIMES}} = \{10, 11, 101, 111, 1011, 1101, \ldots\}$$

$$= \{\text{rep}_2(n) \mid n \text{ is a prime}\}.$$

This problem is decidable. There exists quite elaborate primality testing algorithms. A basic one is to test all the possible divisors up to $\sqrt{n}$. Hence in a finite number of steps, we can decide whether or not $n$ is prime.

As another example involving two parameters, consider the following question. Let $m, n$ be integers such that $m > n > 0$. Is $n$ a square modulo $m$? Does there exist $y$ such that $y^2 \equiv n \pmod{m}$? Again, we consider the coding given by base-2 expansions, but we need an alphabet such as $A = \{(; ); 0; 1; , \}$ to encode pairs $(n, m)$. We list some elements in the language of codings of positive instances

$$\{(1, 10), (1, 11), (1, 100), (1, 101), (100, 101), (1, 110), (11, 110), (100, 110), \ldots\}.$$

It is pretty easy to devise an algorithm to decide this problem.

Let us consider an example coming from graph theory: given a finite directed graph, decide whether this graph is Eulerian or not. Using lemma 3.40, it is easy to check whether a connected graph is Eulerian or not.

EXAMPLE 3.15 (Problems about regular languages).– Many problems about finite automata or regular languages are decidable. Given a DFA $\mathcal{A}$, decide whether or not the language $L(\mathcal{A})$ is empty. Given a DFA $\mathcal{A}$ over $A$, decide whether or not $L(\mathcal{A}) = A^*$. Given two regular languages $L$ and $M$, is $L \cap M = \emptyset$? Is $L \subseteq M$? Is $L = M$? Proofs of these results are left as exercises.

EXAMPLE 3.16 (Problems about logical theories).– The most important problem for the next section arises when we consider a general decision problem concerning first-order theories. Let $\mathfrak{s}$ be a structure with domain $D$ and symbol set $S$. Let $\varphi$ be a sentence over $A_S$. Do we have $\mathfrak{s} \models \varphi$? For a given symbol set $S$, we have defined an alphabet $A_S$ and sentences are well-formed words over $A_S$. Hence, we have a "natural" coding for every instance of this decision problem. Of course, the decidability depends on the considered structure. Theorem 3.24 and Büchi's theorem presented in section 3.4 provide two important examples of decidable theories.

EXAMPLE 3.17.– There are famous undecidable problems like the halting problem for Turing machines and Hilbert's tenth problem[7]. Let us also mention that the post correspondence problem (PCP) is undecidable. This problem related to morphic words was mentioned in section 1.4, Volume 1.

---

7 Let $P \in \mathbb{Z}[X_1, \ldots, X_n]$ be a polynomial with several variables. Does $P$ have an integer root?

## 3.3. Quantifier elimination in Presburger arithmetic

In this chapter, we will discuss sets of integers and more precisely on $k$-recognizable sets of integers. Therefore, structures, such as $\langle \mathbb{N}, + \rangle$ or $\langle \mathbb{N}, +, \leq, 0, 1 \rangle$, play an important role. Recall that we always assume that the equality relation belongs to the structure, so we do not write it explicitly. In the next section, to state Büchi's theorem, we will extend the structure $\langle \mathbb{N}, + \rangle$ with an extra function, namely $V_k$.

### 3.3.1. *Equivalent structures*

The first-order theory of $\mathfrak{p} = \langle \mathbb{N}, + \rangle$, also called *Presburger arithmetic*, is decidable[8]. From the previous sections, we know the meaning of such a statement precisely. There is an algorithm that, given any sentence $\varphi$ from $\mathfrak{p}$, tells us whether or not $\mathfrak{p} \models \varphi$.

To get such a decidability result, we will add some constants and relations to the structure and then proceed to what is called quantifier elimination (see theorem 3.24). Another quite independent technique that will be used in the next sections relies on automata theory (see, for instance, corollary 3.38).

REMARK 3.18.– Note that the binary relation $x \leq y$ can be defined in $\langle \mathbb{N}, + \rangle$ by $(\exists z)(y = x + z)$ because we are working with the domain $\mathbb{N}$. So, we can use $\leq$ as a shortcut for the latter formula. Similarly, $x < y$ can be defined by $(x \leq y) \wedge \neg(x = y)$.

Enriching the structure with the relation $\leq$ permits avoiding the use of some existential quantifier. This observation is quite important and will be repeated several

---

8 See [PRE 91] for a translation of the original paper from 1929.

times in this section because our aim is to build equivalent[9] quantifier-free formulas. Thus, we will be left with a sentence, involving only a finite number of relations (equalities, inequalities and congruence relations) about integers, that has to be checked. Without introducing these relations within the structure, we will not be able to get quantifier-free formulas.

REMARK 3.19.– Every non-negative integer constant is definable in $\langle \mathbb{N}, + \rangle$, more precisely we build a sentence asserting that some variable is equal to a constant. Indeed, $x = 0$ is equivalent to $(\forall y)(x \leq y)$, meaning that $x$ is the smallest non-negative integer. Then, the *successor* function that maps $n$ to $n + 1$ is defined by $y = \text{succ}(x)$ if and only if

$$(x < y) \wedge (\forall z)((x < z) \rightarrow (y \leq z)).$$

This means that $y$ is the least integer greater than $x$. Hence, if $\ell$ is an integer constant and $t$ is a term, then we write $t + \ell$ as a shortcut for $\text{succ}^\ell(t)$.

As an example, $x = 1$ is $\text{succ}(0)$ which is the following formula where $z$ represents zero:

$$(\exists z)((\forall y)(z \leq y) \wedge (z < x) \wedge (\forall y)((z < y) \rightarrow (x \leq y))).$$

Again enriching the structure with the constants $0$ and $1$ permits avoiding the use of some extra quantifiers such as those occurring in the above formula. Indeed, if $1$ is not present in the structure, then it can be defined as above but extra quantifiers are needed. Also, the constant $1$ is enough to define any positive constant, e.g. $2 = 1 + 1$.

REMARK 3.20.– Multiplication by an integer constant is definable in $\langle \mathbb{N}, + \rangle$. Let $t$ be a term. We write $2\,t$ as a shortcut

---

9 Recall that two formulas $\varphi$ and $\psi$ are *equivalent* if they define the same set, i.e. $M_{s,\varphi} = M_{s,\psi}$.

for $t + t$ and more generally, we can define multiplication by a constant by repeating the same term a convenient number of times.

In particular, we can define every congruence relation modulo $m$ (where $m \geq 2$ is fixed) in $\langle \mathbb{N}, + \rangle$. Indeed, two integers $x, y$ are such that $x \equiv y \pmod{m}$ if

$$(\exists q)(\exists q')(\exists r)(x = qm + r \wedge y = q'm + r).$$

Note that $m$ is a constant, and thus multiplication by $m$ is well-defined. We will usually write $\equiv_m$.

As a result of the previous three remarks, we get the following result (equivalent structures were introduced in definition 3.10).

PROPOSITION 3.21.– The structures

$$\langle \mathbb{N}, + \rangle, \ \langle \mathbb{N}, +, \leq, 0, 1 \rangle \ \text{ and } \ \mathfrak{p} = \langle \mathbb{N}, +, \leq, 0, 1, (\equiv_m)_{m \geq 2} \rangle$$

are equivalent.

PROOF.– It is clear that every formula in $\langle \mathbb{N}, + \rangle$ is also a formula in $\langle \mathbb{N}, +, \leq, 0, 1 \rangle$ or $\langle \mathbb{N}, +, \leq, 0, 1, (\equiv_m)_{m \geq 2} \rangle$. For the converse, we use the previous remarks. ∎

At this stage, the readers may wonder why we should consider $\langle \mathbb{N}, +, \leq, 0, 1 \rangle$ or even the structure $\langle \mathbb{N}, +, \leq, 0, 1, (\equiv_m)_{m \geq 2} \rangle$ (with infinitely many relations) instead of a somehow "minimal" structure $\langle \mathbb{N}, + \rangle$. Indeed, we could define exactly the same sets. Whenever the constants 0 and 1 or the relations $\leq$ or $\equiv_m$ are given, then we avoid the use of some extra quantifiers. This is important because our aim is to build quantifier-free formulas[10].

---

10 The number of nested quantifiers gives a measure of complexity for first-order formulas.

EXAMPLE 3.22.– Let us consider a few sentences in the Presburger arithmetic $\mathfrak{p} = \langle \mathbb{N}, +, \leq, 0, 1, (\equiv_m)_{m \geq 2} \rangle$:

$$\varphi \equiv (\forall y)(\exists x)(y = x + x), \quad \psi \equiv (\forall x)(x \equiv_2 0 \vee x \equiv_2 1).$$

We have $\mathfrak{p} \not\models \varphi$ because not every integer is even and $\mathfrak{p} \models \psi$ because every integer is either even or odd. Using the extra constants embedded within the structure (and 3 is a shortcut for $1+1+1$), $\nu \equiv 1+1 = 3$ is also a sentence for which obviously $\mathfrak{p} \not\models \nu$. Similarly, $\tau \equiv 1 < 2$ is also a sentence and $\mathfrak{p} \models \tau$.

REMARK 3.23.– Again, we repeat the reason why we consider an extension of $\langle \mathbb{N}, + \rangle$. Note that $1 + 1 = 3$ is a quantifier-free sentence in $\mathfrak{p} = \langle \mathbb{N}, +, \leq, 0, 1 \rangle$ because we have 1 as a constant embedded within the structure. If we consider only the restricted structure $\langle \mathbb{N}, + \rangle$, then the same sentence already contains quantifiers in the definition of 1, and thus cannot be quantifier-free.

### 3.3.2. *Presburger's theorem and quantifier elimination*

One technique showing the decidability of the first-order theory of $\mathfrak{p}$ is to prove that quantifier elimination applies for Presburger arithmetic: every sentence is equivalent (within $\mathfrak{p}$) to a sentence without quantifiers (such as the last two examples with the sentences $\nu$ and $\tau$). Indeed, in quantifier-free sentences there is no variable left, and thus such a sentence contains only constants and relations.

Now, we are ready to state Presburger's theorem. Be aware that we are considering a structure equivalent to $\langle \mathbb{N}, + \rangle$ but extended with extra relations $\equiv_m$, for all $m \geq 2$, $\leq$ and the constants 0 and 1.

THEOREM 3.24 (Presburger's theorem).– The first-order theory of the structure $\mathfrak{p} = \langle \mathbb{N}, +, \leq, 0, 1, \equiv_m \rangle$ admits an effective quantifier elimination. More precisely, any formula is equivalent to a formula that can be obtained algorithmically and which is a Boolean combination of formulas of the form $s \equiv_m t$, $s \leq t$ or $s = t$. In particular, Presburger arithmetic is decidable.

PROOF[11].– Let $\varphi$ be a sentence in $\mathfrak{p}$. Since $(\forall x)\Psi$ is equivalent to $\neg(\exists x)\neg\Psi$, we can assume that any quantifier appearing in $\varphi$ is an existential one. We can also restrict ourselves to the use of $\wedge$, $\vee$ and $\neg$ because the other logical connectives can be defined from these ones (without introducing any new quantifier). Similarly, since $x = y$ is equivalent to $(x \leq y) \wedge (y \leq x)$, we can assume that any atomic formula is either of the form $t_1 \leq t_2$ or $t_1 \equiv_m t_2$. In such atomic formulas, by the usual rule of cancellation, we can moreover assume that each variable occurring in $t_1$ or $t_2$ occurs either on the left-hand side or on the right-hand side but not on both sides.

We can proceed recursively, and it is enough to deal with the elimination of the innermost quantifier. Assume that we have the formula $(\exists y)\varphi(x_1, \ldots, x_n, y)$ where $\varphi$ does not contain any quantifier and $x_1, \ldots, x_n$ are free with respect to $\varphi$.

**Step one** (restriction to conjunctions of atomic formulas).

Inside a quantifier-free formula, we can get rid of any negation. If we have a formula with several terms, like $\varphi \equiv \rho_1 \vee \neg(\rho_2 \vee (\rho_3 \wedge \rho_4))$, we apply de Morgan's laws possibly recursively. Hence, we may assume that negation, if any, is only applied to atomic formulas, i.e. we get $\rho_1 \vee (\neg\rho_2 \wedge \neg(\rho_3 \wedge \rho_4))$ and finally $\rho_1 \vee (\neg\rho_2 \wedge (\neg\rho_3 \vee \neg\rho_4))$.

---

11 We follow the lines of [END 01, LAT 06] and we have borrowed the running example from [LAT 06]. Note that Presburger's theorem is a direct consequence of Büchi's theorem (theorem 3.35). So the readers can pass the present proof and wait for the proof of Büchi's theorem that is based on automata. Nevertheless, it is worth knowing how to deal with quantifier elimination.

Let $t_1, t_2$ be two terms. The formula $\neg(t_1 \leq t_2)$ is equivalent to $t_2 + 1 \leq t_1$ and $\neg(t_1 \equiv_m t_2)$ is equivalent to $\bigvee_{j \in [\![1, m-1]\!]}(t_1 \equiv_m t_2 + j)$.

Applying de Morgan's laws again (about distributivity), we can assume that $\varphi$ is a disjunction of conjunctions of atomic formulas. As an example, $\rho_1 \vee (\rho_2 \wedge (\rho_3 \vee \rho_4))$ can be replaced with the disjunction $\rho_1 \vee (\rho_2 \wedge \rho_3) \vee (\rho_2 \wedge \rho_4)$.

Since $(\exists y)(\Psi \vee \tau)$ is equivalent to $((\exists y)\Psi) \vee ((\exists y)\tau)$, it is enough to explain how to eliminate an existential quantifier $(\exists y)$ in a conjunction $\varphi$ of atomic formulas. Without loss of generality, we may assume that $y$ appears in each atomic formula (indeed, otherwise there is no constraint on $y$). As an example, consider the formula $(\exists y)\varphi(x_1, x_2, y)$ given by

$$(\exists y)(0 \leq y \wedge x_1 + x_2 \leq 2y + 3 \wedge x_1 + 1 \leq 2x_2 + 3y$$

$$\wedge y \leq x_1 \wedge x_1 + x_2 + 1 \equiv_3 2y).$$

**Step two** (uniformization of the coefficients of $y$).

Let $M$ be the least common multiple of the coefficients of $y$ occurring in $\varphi$. We multiply each inequation by a convenient constant in such a way that in every term, the coefficient of $y$ has been replaced by $M$. Note that such a modification can also be applied to the congruence relations. Let $k \neq 0$. Indeed, $t_1 \equiv_m t_2$ if and only if $kt_1 \equiv_{km} kt_2$. With the running example, we get

$$(\exists y)(0 \leq 6y \wedge 3x_1 + 3x_2 \leq 6y + 9 \wedge 2x_1 + 2 \leq 4x_2 + 6y$$

$$\wedge 6y \leq 6x_1 \wedge 3x_1 + 3x_2 + 3 \equiv_9 6y).$$

**Step three** (elimination of the coefficient of $y$).

We replace $My$ with a new variable $y'$, and we add to the conjunction the atomic formula $y' \equiv_M 0$. Hence, we get a conjunction of atomic formulas of the form $t \leq y' + s$, $y' + s \leq t$ or $y' + s \equiv_m t$ where $s, t$ are terms in which $y'$ does not occur.

For the sake of simplicity, we will write $t - s \leq y'$, $y' \leq t - s$ or $y' \equiv_m t - s$. Hence, we get a formula with three kinds of constraints which is of the form

$$(\exists y') \left( \bigwedge_{j \in [\![1,p]\!]} t_j - s_j \leq y' \wedge \bigwedge_{j \in [\![p+1,p+q]\!]} y' \leq t_j - s_j \right.$$

$$\left. \wedge \bigwedge_{j \in [\![p+q+1,p+q+r]\!]} y' \equiv_{m_j} t_j - s_j \right).$$

The formula is satisfied if there exists $y'$ between some bounds given by the first two conjunctions and $y'$ satisfies some congruence relations.

Let $P$ be the least common multiple of the $m_j$'s occurring in the latter formula. Note that $y' \equiv_{m_j} t_j - s_j$ if and only if $y' + P \equiv_{m_j} t_j - s_j$ for all $j \in [\![p+q+1, p+q+r]\!]$. Hence, we can make use of this periodicity to test a finite number of candidates for $y'$. Indeed, if $\alpha_j$ is a lower bound given by a condition of the form $\alpha_j := t_j - s_j \leq y'$ (we have $p$ of them and if $p = 0$, then the lower bound is 0), then we have only to test all the constraints for

$$\alpha_j, \alpha_j + 1, \ldots, \alpha_j + P - 1.$$

Our running example is

$$(\exists y')(0 \leq y' \wedge 3x_1 + 3x_2 - 9 \leq y' \wedge 2x_1 + 2 - 4x_2 \leq y'$$

$$\wedge \, y' \leq 6x_1$$

$$\wedge \, y' \equiv_9 3x_1 + 3x_2 + 3 \wedge y' \equiv_6 0).$$

The least common multiple of 9 and 6 is 18, and we have three possible lower bounds that are 0, $3x_1 + 3x_2 - 9$ and $2x_1 + 2 - 4x_2$ (since $x_1$ and $x_2$ are still out of the scope of any quantifier, we cannot determine yet which one is the largest). Hence, the original formula $(\exists y)\varphi(x_1, x_2, y)$ is satisfied if and only if, either (replacing $y'$ with $k$)

$$\bigvee_{k \in [\![0,17]\!]} (3x_1 + 3x_2 - 9 \leq k \wedge 2x_1 + 2 - 4x_2 \leq k$$

$$\wedge\, k \leq 6x_1$$

$$\wedge\, k \equiv_9 3x_1 + 3x_2 + 3 \wedge k \equiv_6 0)$$

or (replacing $y'$ with $3x_1 + 3x_2 - 9 + k$)

$$\bigvee_{k \in [\![0,17]\!]} (0 \leq 3x_1 + 3x_2 - 9 + k \wedge 2x_1 + 2 - 4x_2$$

$$\leq 3x_1 + 3x_2 - 9 + k \wedge 3x_1 + 3x_2 - 9 + k \leq 6x_1$$

$$\wedge\, 3x_1 + 3x_2 - 9 + k \equiv_9 3x_1 + 3x_2 + 3 \wedge 3x_1 + 3x_2 - 9 + k \equiv_6 0)$$

or finally (replacing $y'$ with $2x_1 + 2 - 4x_2 + k$)

$$\bigvee_{k \in [\![0,17]\!]} (0 \leq 2x_1 + 2 - 4x_2 + k \wedge 3x_1 + 3x_2 - 9$$

$$\leq 2x_1 + 2 - 4x_2 + k \wedge 2x_1 + 2 - 4x_2 + k \leq 6x_1$$

$$\wedge\, 2x_1 + 2 - 4x_2 + k \equiv_9 3x_1 + 3x_2 + 3 \wedge 2x_1 + 2 - 4x_2 + k \equiv_6 0).$$

So, the resulting formula is a disjunction of these three quantifier-free formulas. ∎

### 3.3.3. *Some consequences of Presburger's theorem*

The first consequence of Presburger's theorem is the decidability of Presburger arithmetic. Then, we discuss the structure of subsets of $\mathbb{N}^n$ defined by a first-order formula in p.

COROLLARY 3.25.– Given a formula $\varphi(x_1, \ldots, x_n)$ in Presburger arithmetic with some free variables $x_1, \ldots, x_n$, we can decide whether or not $\varphi$ is satisfiable.

PROOF.– Recall that $\varphi$ is satisfiable if and only if

$$\mathfrak{p} \models (\exists d_1) \cdots (\exists d_n)\varphi(d_1, \ldots, d_n).$$

Note that using these existential quantifiers, we have a sentence and we can, therefore, apply Presburger's theorem that amounts to testing a finite number of relations about integers. ∎

Let us consider a non-trivial example[12] of provable statement in Presburger arithmetic.

Let $x_1, x_2, \ldots, x_k$ be positive integers. It is well known that every sufficiently large integer can be written as a non-negative integer linear combination of the $x_i$ if and only if $\gcd(x_1, x_2, \ldots, x_k) = 1$. The *Frobenius problem* [KAO 08] is the following: given positive integers $x_1, x_2, \ldots, x_k$ with $\gcd(x_1, x_2, \ldots, x_k) = 1$, find the largest positive integer $g(x_1, x_2, \ldots, x_k)$ that cannot be represented as a non-negative integer linear combination of the $x_i$.

THEOREM 3.26 (chicken McNuggets theorem [VAR 91]).– Suppose that chicken McNuggets can be purchased at McDonald's only in quantities of 6, 9 or 20 pieces. The largest number of McNuggets that cannot be purchased is $g(6, 9, 20) = 43$.

PROOF (UP TO IMPLEMENTATION).– It is a direct consequence of the decidability of Presburger arithmetic. Simply, consider the sentence

$$(\forall n)(n > 43 \to (\exists x, y, z \geq 0)(n = 6x + 9y + 20z))$$

$$\wedge \neg((\exists x, y, z \geq 0)(43 = 6x + 9y + 20z)).$$

Of course, this requires some implementation work to really get a proof and know that $g(6, 9, 20) = 43$. ∎

---

12 This nice example was suggested by J. Shallit. You can also have some fun with the book [VAR 91].

COROLLARY 3.27.– Let $X \subseteq \mathbb{N}$ be a subset of integers. There exists a formula $\varphi(x)$ in Presburger arithmetic with one free variable $x$ such that $X = M_{p,\varphi}$ if and only if $X$ is ultimately periodic.

PROOF.– If $X$ is ultimately periodic, then $X$ is the union of a finite set and finitely many arithmetic progressions. It is, therefore, of the form

$$\{c_1, \ldots, c_k\} \cup d_1 + p\mathbb{N} \cup \cdots \cup d_r + p\mathbb{N}$$

for some integers $p, c_1, \ldots, c_k, d_1, \ldots, d_r$ that can be chosen such that there exists $N$ such that $c_1, \ldots, c_k < N$ and $d_1, \ldots, d_r \geq N$. Hence, $X$ can be described by a formula $\varphi(x)$ of the form

$$(x = c_1) \vee \cdots \vee (x = c_k) \vee ((x \geq N) \wedge ((x \equiv_p d_1) \vee \cdots \vee (x \equiv_p d_r))).$$

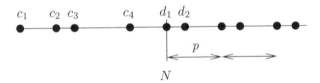

**Figure 3.1.** *An ultimately periodic set*

The converse follows immediately from Presburger's theorem and the particular form of the quantifier-free formula that is obtained.    ∎

The above corollary has a natural expression in a multidimensional setting. Just like for infinite words, apart from a finite region close to the origin, the space can be divided in a finite number of regions and within each of these regions, the set definable in $\langle \mathbb{N}, + \rangle$ has a *periodic pattern*, i.e. there exists a finite hypercube, e.g. a square in dimension 2, that is repeated by translation in all directions parallel to the axes. This follows immediately from the fact that such a set is defined by a Boolean combination of formulas of the form

$s \equiv_m t$, $s \leq t$ or $s = t$. Formally, $X \subseteq \mathbb{N}^n$ has a periodic pattern if there exists a finite set $F \subseteq [\![0, \ell - 1]\!]^n$ such that

$$X = F + \ell\, e_1\, \mathbb{N} + \cdots + \ell\, e_n\, \mathbb{N}.$$

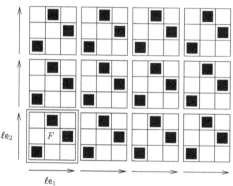

**Figure 3.2.** *A periodic pattern with* $F \subseteq [\![0, 2]\!]^2$ *and* $\ell = 3$

COROLLARY 3.28.– Let $n \geq 2$. Let $X \subseteq \mathbb{N}^n$ be a set in $\mathbb{N}^n$. There exists a formula $\varphi(x_1, \ldots, x_n)$ in Presburger arithmetic with $n$ free variables $x_1, \ldots, x_n$ such that $X = M_{\mathfrak{p}, \varphi}$ if and only if there exists $D$ such that

$$X \cap \{(x_1, \ldots, x_n) \in \mathbb{N}^n \mid \max_i x_i \geq D\}$$

is a finite union of regions delimited by hyperplanes and within each such region a periodic pattern appears.

PROOF.– To give the readers an idea about the proof, we will only consider an example. From Presburger's theorem, we have only to consider particular Boolean combinations of formulas. Take, for instance, the formula

$$\varphi(x_1, x_2) \equiv \rho_1(x_1, x_2) \lor \rho_2(x_1, x_2) \lor \rho_3(x_1, x_2) \lor \rho_4(x_1, x_2) \lor \phi(x_1, x_2)$$

**where**

$$\rho_1(x_1, x_2) \equiv (2x_2 < x_1) \wedge (x_1 + x_2 \equiv_3 0),$$
$$\rho_2(x_1, x_2) \equiv (2x_2 \geq x_1) \wedge (x_2 < x_1) \wedge (x_1 \equiv_4 1),$$
$$\rho_3(x_1, x_2) \equiv (x_2 > x_1) \wedge (x_2 < 3x_1) \wedge ((2x_1 + x_2 \equiv_3 1) \vee (x_1 + x_2 \equiv_3 0)),$$
$$\rho_4(x_1, x_2) \equiv (x_2 \geq 3x_1) \wedge (x_1 \geq 2),$$
$$\phi(x_1, x_2) \equiv (x_1 = 0 \wedge x_2 = 4) \vee (x_1 = 2 \wedge x_2 = 2) \vee (x_1 = 4 \wedge x_2 = 0)$$
$$\vee (x_1 = 5 \wedge x_2 = 0).$$

Each formula $\rho_i$ defines a region delimited by linear equations, and the congruence relations define some pattern that is repeated periodically within this region. The formula $\phi$ adds a finite number of extra points (they are depicted in gray in Figure 3.3). From the expression of $\phi$, the periodic description within each region is satisfied whenever $\max\{x_1, x_2\} \geq 6$. So, in a bounded region near the origin (explaining the constant $D$ in the statement), we can have some arbitrary behavior described by some finite formula. A small portion of the set $M_{\mathsf{p},\varphi}$ is depicted in Figure 3.3. ∎

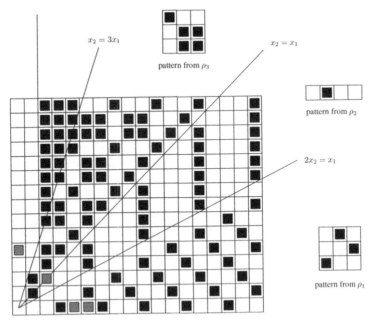

**Figure 3.3.** *A set definable in Presburger arithmetic*

REMARK 3.29.– We can say that a set $X$ of $\mathbb{N}^n$ (or $\mathbb{Z}^n$) is *periodic* if there exists $v \in \mathbb{N}^n$ such that $x \in X$ if and only if $x + v \in X$. As reflected by the next proposition, an alternative generalization of ultimately periodic words is to consider sets of $\mathbb{N}^n$ that are definable in Presburger arithmetic. See, for instance, [DUR 13b] or the discussion about Nivat's conjecture at the end of section 2.8, Volume 1.

EXERCISE 3.3.1.– With the above definitions, show that if $X \subseteq \mathbb{N}^n$ has a periodic pattern then it is periodic. Give a counter-example showing that the converse does not hold.

DEFINITION 3.30.– A set $X$ of $\mathbb{N}^n$ is *linear* if there exists $v_0, v_1, \ldots, v_k \in \mathbb{N}^n$ such that

$$X = v_0 + \mathbb{N}v_1 + \cdots + \mathbb{N}v_k.$$

The vectors $v_1, \ldots, v_k$ are usually called the *periods* of $X$. A set $X$ of $\mathbb{N}^n$ is *semi-linear* if it is a finite union of linear sets[13].

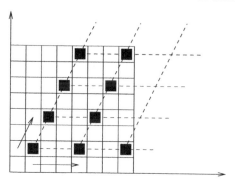

**Figure 3.4.** *A linear set*

EXAMPLE 3.31.– In Figure 3.4, a linear set $v_0 + \mathbb{N}v_1 + \mathbb{N}v_2$ of $\mathbb{N}^2$ is depicted with a constant $v_0 = (1,1)$ and two periods $v_1 = (3,0)$ and $v_2 = (1,2)$.

---

13 We can prove that every semi-linear set is a finite union of linear sets, each of which has linearly independent periods [ITO 69] and [GIN 64].

PROPOSITION 3.32.– [GIN 64] A set $X$ of $\mathbb{N}^n$ is definable in Presburger arithmetic if and only if it is semi-linear.

It is nice that Presburger arithmetic is decidable but so far, we have not mentioned the resources that are needed to effectively test the status of a given sentence. We have not discussed complexity classes so far but, for the readers having some knowledge about complexity theory, let us mention that Presburger arithmetic is beyond NP. We mention the following result that tells us that whatever the algorithm is, there are always sentences in $\mathfrak{p}$ that need (double) exponentially many computational steps to decide [FIS 74].

THEOREM 3.33 (Fischer–Rabin theorem).– There exists a constant $c > 0$ such that for every decision procedure (algorithm) A for Presburger arithmetic $\mathfrak{p}$, there exists an integer $N$ so that for every $n > N$ there exists a sentence $\varphi$ of length $n$ for which A requires more than $2^{2^{cn}}$ computational steps to decide whether $\mathfrak{p} \models \varphi$. This statement applies also in the case of non-deterministic algorithms.

The previous theorem implies that not only do algorithms require a super-exponential number of computational steps in the worst-case, but also that proofs of true statements concerning addition of natural numbers can be super-exponentially long. Otherwise stated, for any algorithm that we could devise, even though this algorithm behaves well (i.e. tests can be performed quickly) in some instances, there always exists an instance for which that particular decision algorithm will require a prohibitive number of steps. For more recent and accurate results, see the bibliographic notes.

### 3.4. Büchi's theorem

In this section, after introducing the notion of $k$-definable set of integers, we show that a set is $k$-definable if and only if it is $k$-recognizable (see definition 1.35). The most important thing for us is to replace a formula defining a set of $\mathbb{N}^n$ with a DFA recognizing base-$k$ expansions of elements belonging to that set. In the second part of this section, we generalize this result to Pisot numeration systems.

### 3.4.1. *Definable sets*

In remark 1.18, Volume 1, we introduced, for a prime $p$, the $p$-adic valuation $\nu_p$. Let $k \geq 2$ be an integer. We define a related map $V_k : \mathbb{N} \to \mathbb{N}$. Let $x \in \mathbb{N}_{\geq 1}$. We write $x$ as $k^n q$ where $n \geq 0$ and $q$ is an integer not divisible by $k$ (such a decomposition is unique) and we set $V_k(x) = k^n$. Otherwise stated, $V_k(x)$ is the largest power of $k$ dividing $x$. Moreover, we set $V_k(0) = 1$. In particular, if $k$ is a prime, then we have the following relation with the $k$-adic valuation: $V_k(x) = k^{\nu_k(x)}$.

DEFINITION 3.34.– Let $n \geq 1$. We will consider the structure $\mathfrak{n}_k = \langle \mathbb{N}, +, V_k \rangle$, and therefore define subsets of $\mathbb{N}^n$ as in [3.3]. A subset $X \subseteq \mathbb{N}^n$ is $k$-*definable* if there exists a formula $\varphi = \varphi(x_1, \ldots, x_n)$ in $\mathfrak{n}_k$ with $n$ free variables $x_1, \ldots, x_n$ such that $X = M_{\mathfrak{n}_k, \varphi}$, i.e.

$$X = \{(d_1, \ldots, d_n) \in \mathbb{N}^n \mid \mathfrak{n}_k \models \varphi(d_1, \ldots, d_n)\}.$$

Let us give some examples of $k$-definable sets. First, we give some examples of formulas in $\mathfrak{n}_2$:

$$\varphi_1(x) \equiv (\exists y)(x = y + y),$$

$$\varphi_2(x) \equiv V_2(x) = x,$$

$$\varphi_3(x) \equiv (\exists y)(\exists z)(V_2(y) = y \wedge V_2(z) = z \wedge x = y + z).$$

It is easy to see that these formulas define the following sets $M_{\mathfrak{n}_2, \varphi_1} = 2\mathbb{N}$, $M_{\mathfrak{n}_2, \varphi_2} = \{2^n \mid n \in \mathbb{N}\}$ and $M_{\mathfrak{n}_2, \varphi_3} = \{2^m + 2^n \mid m, n \in \mathbb{N}\}$. Recall that we write $2y$ as a shortcut for $y + y$ and more generally, we can define multiplication by a constant in $n_k$.

Consider the formulas:

$$\varphi_4(x_1, x_2) \equiv x_1 < x_2,$$

$$\varphi_5(x_1, x_2) \equiv (\exists y)(x_1 + x_2 = y + y),$$

$$\varphi_6(x_1, x_2) \equiv (V_2(x_1) = x_1 \wedge x_1 = x_2) \vee (\neg(V_2(x_1) = x_1)$$

$$\wedge x_2 \leq x_1).$$

There is not much to say about $M_{n_2,\varphi_4}$. The set $M_{n_2,\varphi_5}$ consists of points in $\mathbb{N}^2$ whose two components sum up to an even number. A small region of this set is depicted in Figure 3.5 leading to some checkerboard pattern that will repeat in the whole plane. The set $M_{n_2,\varphi_6}$ is a bit more complex to describe (because it involves the extra function $V_2$). If $x_1$ is a power of two, there is only one point in the set whose first coordinate is $x_1$. This point is $(x_1, x_1)$. Otherwise, i.e. if $x_1$ is not a power of two, then all points $(x_1, x_2)$ with $x_2 \leq x_1$ are in the set. In Figure 3.5, powers of two are visually detected with vertical white bars.

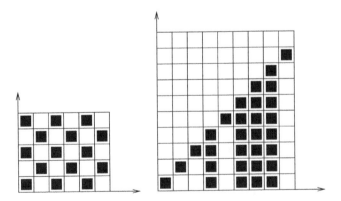

**Figure 3.5.** *Two 2-definable sets*

### 3.4.2. *A constructive proof of Büchi's theorem*

The proof of Büchi's theorem is important because it is constructive: it provides us with an algorithm that replaces any formula with a convenient DFA. See remark 3.36 about practical concerns.

THEOREM 3.35 (Büchi's theorem [BÜC 60]).– Let $k \geq 2$. A set $X \subseteq \mathbb{N}^n$ is $k$-definable if and only if it is $k$-recognizable. Precisely, we can obtain effectively a DFA accepting $\mathrm{rep}_k(X)$ from a formula in $\mathfrak{n}_k$ defining $X \subseteq \mathbb{N}^n$ and conversely.

PROOF[14].– *From formulas to automata.* Assume that $\varphi(x_1, \ldots, x_n)$ is a formula defining $X \subseteq \mathbb{N}^n$, i.e. $M_{\mathfrak{n}_k, \varphi} = X$. We will produce a DFA $\mathcal{A}_\varphi$ reading from right to left $n$-tuples of words over $[\![0, k-1]\!]$ (i.e. least significant digit first) and recognizing the language

$$\{(w_1, \ldots, w_n) \mid (\mathrm{val}_k(w_1), \ldots, \mathrm{val}_k(w_n)) \in X$$

$$\text{and } |w_i| = |w_j|, \ \forall i, j\}.$$

In particular, when considering base-$k$ expansions of $n$ integers, the shortest are padded to the left with leading zeroes[15] to get $n$ words of the same length. Another way to express that the $n$ entries have the same length is to observe that

$$(w_1, \ldots, w_n) \in ([\![0, k-1]\!]^n)^*.$$

Hence, we have a word over a multidimensional alphabet.

We repeat ourselves by stressing on the fact that given a formula $\varphi$ with $n$ free variables, our aim is to get an automaton $\mathcal{A}_\varphi$ reading $n$-tuples of base-$k$ expansions such

---

14 We follow essentially the lines of [BRU 94].
15 This is exactly the convention taken in definition 2.45.

that the integers $d_1, \ldots, d_n$ satisfy the formula $\varphi$, i.e. $n_k \models \varphi(d_1, \ldots, d_n)$, if and only if

$$(0^{m-|\operatorname{rep}_k(d_1)|} \operatorname{rep}_k(d_1), \ldots, 0^{m-|\operatorname{rep}_k(d_n)|} \operatorname{rep}_k(d_n)) \qquad [3.4]$$

is accepted by $\mathcal{A}_\varphi$, where $m = \max_i |\operatorname{rep}_k(d_i)|$. Note that at least one of the words in the above $n$-tuple of base-$k$ expansions starts with a non-zero digit. Recall that automata over a multidimensional alphabet were introduced in section 1.3. It is clear that a language $L \subseteq (\llbracket 0, k - 1 \rrbracket^n)^*$ is regular if and only if $(0, \ldots, 0)^* L$ is regular. So, in every construction, we will consider automata accepting not only words of the type [3.4] but also all the words

$$(0^{j-|\operatorname{rep}_k(d_1)|} \operatorname{rep}_k(d_1), \ldots, 0^{j-|\operatorname{rep}_k(d_n)|} \operatorname{rep}_k(d_n))$$

for all $j \geq \max_i |\operatorname{rep}_k(d_i)|$. Indeed, when complementing a DFA, it is important that all the representations of a $n$-tuple of integers share the same acceptance status.

We proceed by induction on the formula. In $\langle \mathbb{N}, +, V_k \rangle$, atomic formulas are of one of the three following forms $x = y$, $y = V_k(x)$ or $z = x + y$. Considering a structure with a function $f$, we obviously get an equivalent structure by replacing $f$ with the relation $\hat{f}$ that is the graph of $f$. Hence, we can assume that we have three relations

$$\{(x, y) \mid x = y\} \subset \mathbb{N}^2,$$

$$\{(x, y) \mid y = V_k(x)\} \subset \mathbb{N}^2 \text{ and } \{(x, y, z) \mid z = x + y\} \subset \mathbb{N}^3.$$

These relations are easily seen to be $k$-recognizable sets. The DFA for the last relation has been given in example 1.19, see Figure 1.8 for a DFA reading least significant digit first in the case $k = 2$. A DFA for the relation associated with $V_2$ is given in Figure 3.6 (again reading least significant digit first). Note that another DFA for $V_2$ is given in Figure 1.6. But this latter one reads most significant digit first, and $V_2(0)$ is not taken into account. Over $[0, k - 1]$, the language for the relation $V_k$ is

$$\{(wa0^n, 0^{|w|}10^n) \mid a \in [1, k-1], w \in [0, k-1]^*\}$$

$$\cup \{(0^{n+1}, 0^n 1) \mid n \geq 0\} .$$

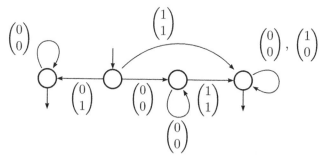

**Figure 3.6.** *A DFA accepting $\{(x, y) \mid y = V_2(x)\}$ in base 2 (least significant digit first)*

Since $\psi \wedge \tau$ is equivalent to $\neg(\neg \psi \vee \neg \tau)$, $(\forall x)\psi$ is equivalent to $\neg((\exists x)\neg\psi)$, $\psi \rightarrow \tau$ is equivalent to $\neg\psi \vee \tau$, it is enough to consider the following formulas $\neg\psi$, $\psi \vee \tau$ and $(\exists x)\psi$ assuming (by the induction hypothesis) that we have at hand automata $\mathcal{A}_\psi$ and $\mathcal{A}_\tau$ recognizing, respectively, $M_{n_k, \psi}$ and $M_{n_k, \tau}$.

Let $\varphi(x_1, \ldots, x_p, y_1, \ldots, y_q, z_1, \ldots, z_r)$ be the formula

$$\psi(x_1, \ldots, x_p, y_1, \ldots, y_q) \vee \tau(x_1, \ldots, x_p, z_1, \ldots, z_r) .$$

Replace the DFA $\mathcal{A}_\psi$ over $[0, k-1]^{p+q}$ with a DFA over $[0, k-1]^{p+q+r}$ where each edge with label $(c_1, \ldots, c_{p+q})$ is replaced with $k^r$ edges with labels

$$(c_1, \ldots, c_{p+q}, s_1, \ldots, s_r), \quad \forall s_1, \ldots, s_r \in [0, k-1] .$$

Replace the DFA $\mathcal{A}_\tau$ over $[0, k-1]^{p+r}$ with a DFA over $[0, k-1]^{p+q+r}$ where each edge with label $(c_1, \ldots, c_p, c_{p+1}, \ldots, c_{p+r})$ is replaced with $k^q$ edges with labels

$$(c_1, \ldots, c_p, s_1, \ldots, s_q, c_{p+1}, \ldots, c_{p+r}), \quad \forall s_1, \ldots, s_q \in [\![0, k-1]\!].$$

We have two automata over the same alphabet, and their union is a convenient non-deterministic finite automaton (NFA)[16] for $\mathcal{A}_{\psi \vee \tau}$.

Let $\psi$ be a formula (with $p$ free variables) and the corresponding DFA $\mathcal{A}_\psi$. We have seen in Chapter 1 that regular languages are closed under complement. Hence, the complement of the DFA $\mathcal{A}_\psi$, where the status of final or non-final of every state is inverted, corresponds exactly to $\neg \psi$. Indeed, this DFA accepts $([\![0, k-1]\!]^p)^* \setminus L(\mathcal{A}_\psi)$. Again, this is the reason why we have considered DFA accepting all (or none) of the base-$k$ expansions of a given $n$-tuple of integers. Because if it was not the case, one $n$-tuple of integers could have some base-$k$ expansions accepted by a DFA and other base-$k$ expansions with a different number of leading zeroes accepted by its complement.

To complete this part of the proof, we consider a formula $\varphi(x_1, \ldots, x_p)$ given by $(\exists x)\psi(x, x_1, \ldots, x_p)$. By the induction hypothesis, we start with a DFA $\mathcal{A}_\psi$ over $[\![0, k-1]\!]^{p+1}$. Replace this DFA $\mathcal{A}_\psi$ with an NFA over $[\![0, k-1]\!]^p$ where each edge with label $(a, c_1, \ldots, c_p)$ is replaced with $(c_1, \ldots, c_p)$. Every accepting path in this NFA labeled by $(w_1, \ldots, w_p)$ corresponds to at least one accepting path in $\mathcal{A}_\psi$ labeled by $(w, w_1, \ldots, w_p)$ for some $w \in [\![0, k-1]\!]^*$ and conversely. Otherwise stated, there exists integers $d_1, \ldots, d_p$ such that $\mathrm{val}_k(w_i) = d_i$ for all $i$, i.e. $\mathfrak{n}_k \models \varphi(d_1, \ldots, d_n)$, if and only if there exists an integer $d$ such that $\mathrm{val}_k(w) = d$ and $\mathfrak{n}_k \models \psi(d, d_1, \ldots, d_n)$. Now the NFA has to be determinized, and we have to allow leading zeroes in the accepted expansions. These construction steps have to be carried on to treat inductively the considered formula.

---

16 We can slightly modify the construction of the product automaton of two DFAs introduced in definition 1.10 to obtain a DFA accepting the union of two regular languages. Simply take $(F \times Q') \cup (Q \times F')$ as set of final states.

**From automata to formulas**. For this part of the proof, we only discuss the general philosophy and we start with a digression.

A usual way to prove that every computable function is recursive is to proceed as follows. If a function $f : \mathbb{N}^n \to \mathbb{N}$ is computed by a (deterministic) Turing machine $\mathcal{M}$, then we define a convenient coding of the states and the meaningful part of the tape of $\mathcal{M}$. Hence, configurations of the Turing machine are coded as tuples of integers. Several primitive recursive functions are devised to code the input of $\mathcal{M}$ and also its transition map. Therefore, a computation of $\mathcal{M}$ for a given input can be translated into a recursive map: the initial configuration given by the input and the initial state is coded by a tuple of integers, we apply iteratively the map coding the transition function until we get, for the first time, the coding of a halting state.

The idea is quite similar here. Assume that we are given a DFA $\mathcal{A}$ recognizing some subset of $n$-tuples of words $(w_1, \ldots, w_n)$ in $(\llbracket 0, k-1 \rrbracket^n)^*$ that corresponds to a subset of $\mathbb{N}^n$. We have to encode this DFA $\mathcal{A}$ in $\langle \mathbb{N}, +, V_k \rangle$ by a formula $\psi$ with $n$ free variables in such a way that $(w_1, \ldots, w_n)$ is accepted by $\mathcal{A}$ if and only if $\psi(\mathrm{val}_k(w_1), \ldots, \mathrm{val}_k(w_n))$ holds.

If $\mathcal{A}$ has $r$ states, then each state is coded by a $r$-tuple of the type $(0, \ldots, 0, 1, 0, \ldots, 0)$. Thus, a finite sequence of states in $\mathcal{A}$ corresponds to a concatenation component-wise $(v_1, \ldots, v_r) \in (\{0,1\}^r)^*$ of such $r$-tuples, and thus leads to the definition of $r$ integers $(\mathrm{val}_k(v_1), \ldots, \mathrm{val}_k(v_r))$. As an example, assume that we have two states $q_0$ and $q_1$ coded, respectively, by $(1,0)$ and $(0,1)$. The sequence $q_0, q_1, q_0, q_0$ corresponds to $(1011, 0100)$. If $k = 3$, then the sequence is coded by the pair of integers $(31, 9)$. Note that the coding is of course injective but not surjective.

An accepting path (for a given input) is a sequence of states that has to start in the initial state, follows the transition function of $\mathcal{A}$ and ends in a final state. All these conditions can be translated[17] into a formula

$$\varphi(x_1, \ldots, x_n, y_1, \ldots, y_r)$$

which evaluates to true if and only if $y_1, \ldots, y_r$ are integers coding an accepting path in $\mathcal{A}$ whose transitions are those given by reading the $k$-ary expansions of $x_1, \ldots, x_n$ (with the usual convention of padding shortest expansions with zeroes). Thus, the $n$-tuple $(w_1, \ldots, w_n)$ in $(\llbracket 0, k - 1 \rrbracket^n)^*$ is recognized by $\mathcal{A}$ if and only if the formula

$$(\exists y_1) \cdots (\exists y_r) \varphi(\mathrm{val}_k(w_1), \ldots, \mathrm{val}_k(w_n), y_1, \ldots, y_r))$$

is satisfied. Hence, derived from $\mathcal{A}$, we get the definable set

$$\{(x_1, \ldots, x_n) \in \mathbb{N}^n \mid \mathbf{n}_k \models (\exists y_1) \cdots (\exists y_r) \varphi(x_1, \ldots, x_n, y_1, \ldots, y_r)\} \, .$$

∎

REMARK 3.36.– The proof of Büchi's theorem provides us with an algorithm that replaces any formula with a convenient DFA. We start with basic automata for equality, $V_k$ and addition and build more complex automata from them. The construction of these automata follows the construction of the formula. Note that the presence of an existential or universal quantifier leads to a non-deterministic automaton that needs to be determinized. Recall that determinization can lead to an exponential blow up of the number of states: if an NFA has $n$ states, then an equivalent DFA built using the subset construction can have up to $2^n$ states. Hence, the approach provided by Büchi's theorem seems challenging: nested quantifiers could, in the

---

17 The details about how to get such a formula can be found in [BRU 94]. For instance, if we have a sequence of $\ell$ states, then the coefficients of $k^{\ell-1}$ in the base-$k$ expansion of $y_1, \ldots, y_r$ have to match the coding of the initial state. All the conditions can be translated in convenient properties about base-$k$ expansions.

worst-case, lead to a tower of exponentials (see the bibliographic notes for complexity issues). Nevertheless, it seems that luckily, we can handle many interesting problems in combinatorics on words (see section 3.5).

We now review some important corollaries of Büchi's theorem. Then to conclude with this section, we will explain how Büchi's theorem can be extended to Pisot numeration systems like the Zeckendorf system built on the Fibonacci sequence. The first corollary is a restatement of proposition 1.39.

COROLLARY 3.37.– An infinite word $\mathbf{w} = w_0 w_1 w_2 \cdots \in A^{\mathbb{N}}$ is $k$-automatic if and only if, for all $a \in A$, $\mathrm{fiber}_a(\mathbf{w}) = \{i \in \mathbb{N} \mid w_i = a\}$ is $k$-definable.

PROOF.– Let $a \in A$. We apply Cobham's theorem on automatic sequences, theorem 2.28, Volume 1. If $\mathbf{w}$ is $k$-automatic, there exists a deterministic finite automation with output (DFAO) $\mathcal{A}$ fed with base-$k$ expansions such that, for all $j \geq 0$, $w_j = \mu(\delta(q_0, \mathrm{rep}_k(j)))$ where $\delta$ is the transition function, $q_0$ is the initial state and $\mu : Q \to A$ is the output function of $\mathcal{A}$. From the DFAO $\mathcal{A}$, we define a DFA where a state $q$ is final if and only if $\mu(q) = a$. This DFA accepts exactly $0^* \{\mathrm{rep}_k(i) \mid w_i = a\}$. From Büchi's theorem, $\mathrm{fiber}_a(\mathbf{w})$ is $k$-definable.

Conversely, assume that, for all $a \in A = \{a_1, \ldots, a_p\}$, $\mathrm{fiber}_a(\mathbf{w})$ is $k$-definable. Büchi's theorem gives, for each $j \in [\![1, p]\!]$, a DFA $\mathcal{A}_j$ having $F_j$ as set of final states and recognizing $0^* \{\mathrm{rep}_k(i) \mid w_i = a_j\}$. Consider the product of these DFAs (see definition 1.10). From this product, we define a DFAO producing $\mathbf{w}$ where the output function $\mu$ is given by

$$\mu(q_1, \ldots, q_p) = a_j \Leftrightarrow q_j \in F_j \,.$$

This map is well-defined because the fibers form a partition of $\mathbb{N}$, i.e. any word $w \in [\![0, k-1]\!]^*$ is recognized by a unique $\mathcal{A}_j$. ∎

Now, we extend Presburger's theorem from $\langle \mathbb{N}, + \rangle$ to $\mathfrak{n}_k = \langle \mathbb{N}, +, V_k \rangle$.

COROLLARY 3.38.– The (first-order) theory of $\mathfrak{n}_k = \langle \mathbb{N}, +, V_k \rangle$ is decidable.

PROOF.– Let $\varphi$ be a sentence (i.e. a formula with no free variable). If the formula is a conjunction or disjunction of sentences, then we can treat each clause separately. Without loss of generality, we may assume that $\varphi$ is thus restricted to a unique clause. There is an outermost quantifier appearing in $\varphi$. Replacing $(\forall x)\psi$ with $\neg((\exists x)\neg\psi)$, we may assume that $\varphi$ is of the form $(\exists x)\psi$, and $\psi$ has $x$ as unique free variable. We can easily take care of negation by considering the complement of languages. Following the proof of Büchi's theorem, we get a DFA recognizing the language $L$ made up of the base-$k$ expansions of the elements in $\{n \in \mathbb{N} \mid \mathfrak{n}_k \models \psi(n)\}$. Note that $\psi$ is satisfiable (or equivalently $\varphi$ is true) if and only if $L$ is non-empty. Whether or not a DFA accepts at least one word is easily seen to be decidable. ∎

REMARK 3.39.– This corollary also gives an alternative proof (involving construction of automata) of Presburger's theorem. Indeed, if the theory of $\langle \mathbb{N}, +, V_k \rangle$ is decidable, then the one with $\langle \mathbb{N}, + \rangle$ is also decidable.

We borrow the formulation of Büchi's theorem as stated by Charlier, Rampersad and Shallit [CHA 12]. Indeed, this result is important for the applications in combinatorics on words that we will present.

THEOREM 3.40.– Let $k \geq 2$. If we can express a property of a $k$-automatic sequence x using quantifiers, logical operations, integer variables, the operations of addition, subtraction, indexing into x (i.e. if $n$ is an integer variable, we have access to x($n$)) and comparison of integers or elements of x, then this property is decidable.

PROOF.– Let x be a $k$-automatic sequence. From corollary 3.37, each set $\text{fiber}_a(\mathbf{x})$ is $k$-definable by a formula $\chi_{\mathbf{x},a}(n)$ in $\mathfrak{n}_k$. We can, therefore, express the fact that $\mathbf{x}(i) = \mathbf{x}(j)$ by the formula

$$\text{eq}_{\mathbf{x}}(i,j) \equiv \bigvee_{a \in A} \left( \chi_{\mathbf{x},a}(i) \wedge \chi_{\mathbf{x},a}(j) \right).$$

Now, assume that we have a property about an infinite word that can be expressed by a formula in $\mathfrak{n}_k$. As an example, consider the property that an infinite word w contains arbitrarily long squares. This can be express by the formula

$$(\forall N)(\exists \ell)(\ell > N \wedge (\exists i)(\forall k)(k < \ell \rightarrow \mathbf{w}(i+k) = \mathbf{w}(i + \ell + k))).$$

Indeed, the first two quantifiers translate the fact that we are considering arbitrarily long squares. Then, there exists some position $i$ where a square of length $2\ell$ occurs.

Consider the formula

$$(\forall N)(\exists \ell)(\ell > N \wedge (\exists i)(\forall k)(k < \ell \rightarrow \text{eq}_{\mathbf{x}}(i+k, i + \ell + k))).$$

This is a sentence in $\mathfrak{n}_k$, and we can make use of corollary 3.38: we can decide whether or not the $k$-automatic word x has arbitrarily long squares. ∎

Cobham's theorem given by theorem 1.45 can be restated in terms of $q$-definable sets instead of $q$-recognizable sets. We simply use Büchi's theorem and corollary 3.27.

THEOREM 3.41.– Let $p, q \geq 2$ be two multiplicatively independent integers. If a set $X \subseteq \mathbb{N}$ is both $p$-definable and $q$-definable, then $X$ is definable in $\langle \mathbb{N}, + \rangle$.

This result can be extended to a multidimensional setting: it is the so-called Cobham–Semenov theorem.

THEOREM 3.42.– Let $p, q \geq 2$ be two multiplicatively independent integers. If a set $X \subseteq \mathbb{N}^n$ is both $p$-definable and $q$-definable, then $X$ is definable in $\langle \mathbb{N}, + \rangle$.

### 3.4.3. *Extension to Pisot numeration systems*

Recall that integer base systems are special cases of Pisot numeration systems as described in section 2.4. Let $U$ be a Pisot numeration system. Let us summarize the main properties that are useful within this logical framework:

1) The set $\mathbb{N}$ is $U$-recognizable, i.e. $\mathrm{rep}_U(\mathbb{N})$ is regular.

2) The graph of the normalization function $\nu_{U,B}$ is $U$-recognizable. Hence, the graph of addition is $U$-recognizable.

The key argument for the extension of Büchi's theorem is that the primitive relations for $=$ and $+$ can be represented by finite automata. It is also necessary that $\mathbb{N}$ be $U$-recognizable (such a consideration is not taken into account for the usual integer base system because in that case $0^* \mathrm{rep}_k(\mathbb{N}) = [\![0, k - 1]\!]^*$). We introduce the map $V_U$ where $V_U(n)$ is the smallest $U_i$ appearing in the normal $U$-representation of $n \geq 1$ with a non-zero coefficient. Otherwise stated, if $\mathrm{rep}_U(n) = c_d \cdots c_r 0^r$ with $c_r \neq 0$, then $V_U(n) = U_r$. We set $V_U(0) = 1$.

REMARK 3.43.– If the canonical alphabet $A_U$ of the numeration system is $\{0, 1\}$, then the set $\{(x, y) \mid y = V_U(x)\}$ is $U$-recognizable. We have to consider the intersection of the language accepted by the DFA depicted in Figure 3.6 on page 161 and the language

$$\{(u, v) \mid u \in 0^* \mathrm{rep}_U(\mathbb{N}), v \in \{0, 1\}^*, |u| = |v|\}.$$

Indeed, we must ensure that we are dealing with valid $U$-representations. A similar construction can be achieved for an arbitrary finite alphabet. Hence, the graph of $V_U$ is $U$-recognizable.

We consider the natural extension of $k$-definable sets.

DEFINITION 3.44.– Let $U$ be a Pisot numeration system. We will constantly consider the structure $\mathfrak{n}_U = \langle \mathbb{N}, +, V_U \rangle$ and define subsets of $\mathbb{N}^n$ as in [3.3]. A subset $X \subseteq \mathbb{N}^n$ is *U-definable* if there exists a formula $\varphi = \varphi(x_1, \ldots, x_n)$ in $\mathfrak{n}_U$ with $n$ free variables $x_1, \ldots, x_n$ such that $X = M_{\mathfrak{n}_U, \varphi}$, i.e.

$$X = \{(d_1, \ldots, d_n) \in \mathbb{N}^n \mid \mathfrak{n}_U \models \varphi(d_1, \ldots, d_n)\}.$$

THEOREM 3.45.– [BRU 97] Let $U$ be a Pisot numeration system. A set $X \subseteq \mathbb{N}^n$ is $U$-definable if and only if it is $U$-recognizable. Precisely, we can obtain effectively a DFA accepting $\mathrm{rep}_U(X)$ from a formula in $\mathfrak{n}_U$ defining $X \subseteq \mathbb{N}^n$ and conversely.

PROOF.– *From formulas to automata.* We follow the same procedure as the one described in the proof of Büchi's theorem. Given a formula $\varphi$, we proceed by induction on $\varphi$. In $\langle \mathbb{N}, +, V_U \rangle$, atomic formulas are of one of the three following forms $y = x$, $y = V_U(x)$ or $z = x + y$, and their graphs are $U$-recognizable. See remark 3.43 and corollary 2.62.

The only difference occurs when considering the complement (i.e. the negation of a formula). If a DFA accepts a language $\mathrm{rep}_U(X)$ over $A_U$ for some $X \subseteq \mathbb{N}$, then we have to consider $(A_U^* \setminus L) \cap 0^* \mathrm{rep}_U(\mathbb{N})$ for the set of $U$-representations of $\mathbb{N} \setminus X$. Indeed, intersection with $0^* \mathrm{rep}_U(\mathbb{N})$ is needed to discard non-normal $U$-representations. A similar observation applies for multidimensional alphabets when considering representations of subsets of $\mathbb{N}^n$.

**From automata to formulas**. Again, we can follow the general scheme presented in the proof of Büchi's theorem. Nevertheless, we have to be careful in the coding of a path because paths are coded by words over $A_U$. But not any word over $A_U$ is a normal $U$-representation. Such a refinement was not necessary for integer base numeration systems because every word over $[\![0, k-1]\!]^*$ is a valid base-$k$ expansion. To understand the problem, assume that we are dealing with the Zeckendorf numeration system and that we have two states $q_0$ and $q_1$ coded, respectively, by $(1,0)$ and $(0,1)$. If we have a loop on $q_0$ and we consider a sequence of states where $q_0$ is repeated twice, like $q_0, q_0$, then we get the coding $(11, 00)$. But recall that two consecutive 1's are not allowed in Zeckendorf representations. Hence, 11 is not a valid coding. To get rid of this problem, we can replace the DFA with an equivalent DFA where every cycle has length of at least 2. If we are dealing with another Pisot numeration system, the same kind of trick can be applied: create sufficiently long cycles such that between any two 1's occurring in the coding of a path there are at least $K$ zeroes ensuring valid representations (see exercise 3.4.1). Therefore, in the coding of any path, we will never have two consecutive 1's. We will not discuss these technical details[18] here. See [BRU 97, Lemma 17].    ∎

EXERCISE 3.4.1.– [BRU 97, Lemma 17] Let $U = (U_n)_{n \geq 0}$ be a Pisot numeration system. Prove that there exists a constant $K$ such that, for all $\ell, n \geq 0$ and all $m_1, \ldots, m_\ell \geq K$, the word $10^{m_1} 10^{m_2} 1 \cdots 10^{m_\ell} 10^n$ is a normal $U$-representation. In particular, for the Zeckendorf (respectively, Tribonacci) numeration system, we can choose $K = 1$ (respectively, $K = 2$).

## 3.5. Some applications

We now present some of the benefits that can be gathered from Büchi's theorem.

### 3.5.1. *Properties about automatic sequences*

We can easily obtain proofs for some results stated in Chapter 2, Volume 1, about closure properties of $k$-automatic sequences: periodic decimation, perfect shuffle, finite modifications and image under a uniform morphism. The

---

18 An alternative is to code each state $q_j$ by a $r$-tuple of words $(u_{j,1}, \ldots, u_{j,r})$ in $(\{0, 1\}^r)^*$ in such a way that any concatenation $u_{j_1,s} \cdots u_{j_\ell,s}$ is a valid $U$-representation.

compactness and the expressiveness of the formulas can replace (or at least hide) construction of some intricate automata or morphisms.

Let $c > d \geq 0$. Let $k \geq 2$. In proposition 2.36, Volume 1, we claimed that if the word $\mathbf{w} = (w_n)_{n\geq 0} \in A^{\mathbb{N}}$ is $k$-automatic, then the word $\mathbf{z} = (w_{cn+d})_{n\geq 0}$ is also $k$-automatic.

PROOF OF PROPOSITION 2.36.– For all $a \in A$, the set fiber$_a(\mathbf{w})$ is $k$-definable by a formula $\chi_{\mathbf{w},a}(n)$. Recall that $cn$ is a shortcut for $n + \cdots + n$ (with $c$ terms). Then, for all $a \in A$, fiber$_a(\mathbf{z})$ is $k$-definable by the formula $\chi_{\mathbf{z},a}(n) = \chi_{\mathbf{w},a}(cn+d)$. The conclusion follows from corollary 3.37. ∎

Let $k \geq 2$. Let $\mathbf{w}_0 = (w_{0,n})_{n\geq 0}, \ldots, \mathbf{w}_{r-1} = (w_{r-1,n})_{n\geq 0}$ be $k$-automatic words over $A$. In proposition 2.37, Volume 1, we claimed that the perfect shuffle $\mathbf{u} = (u_n)_{n\geq 0}$ defined by $u_{nr+i} = w_{i,n}$, for all $n \geq 0$, and all $i \in [\![0, r-1]\!]$, is $k$-automatic.

PROOF OF PROPOSITION 2.37.– For all $a \in A$ and all $j \in [\![0, r-1]\!]$, the set fiber$_a(\mathbf{w}_j)$ is $k$-definable by a formula $\chi_{\mathbf{w}_j,a}(n)$. Then, for all $a \in A$, fiber$_a(\mathbf{u})$ is $k$-definable by the formula

$$\chi_{\mathbf{u},a}(n) = (\exists m) \bigvee_{j=0}^{r-1} (n = rm + j \wedge \chi_{\mathbf{w}_j,a}(m)).$$

Note that $r$ is a given constant, hence we have a finite formula in the above description. Indeed, by Euclidean division, there exists a unique $m$ and a unique $j$ such that $n = rm + j$. So, in the disjunction at most one clause can be satisfied. ∎

In proposition 2.38, Volume 1, we claimed the following. Let $k \geq 2$. Let $u$ be a finite word. If $\mathbf{w}$ is a $k$-automatic word, then $\mathbf{z} = u\mathbf{w}$ is also $k$-automatic.

PROOF OF PROPOSITION 2.38.– For all $a \in A$, the set fiber$_a(\mathbf{w})$ is $k$-definable by a formula $\chi_{\mathbf{w},a}(n)$. Then, for all $a \in A$, fiber$_a(\mathbf{z})$ is $k$-definable by the formula

$$\chi_{\mathbf{z},a}(n) = \bigvee_{j:u_j=a} (n = j) \vee (|u| < n \wedge (\exists m)((m + |u| = n) \wedge \chi_{\mathbf{w},a}(m))).$$

The meaning is the following one. The first part of the formula deals with the prefix of length $|u|$, and the second part means that there is a symbol $a$ in $uw$ in position $m + |u|$ if and only if there is a symbol $a$ in w in position $m$. ∎

EXERCISE 3.5.1.– Prove proposition 2.39, Volume 1. Let $k \geq 2$. If w is a $k$-automatic word, then its shift $\sigma(w)$ is also $k$-automatic.

In proposition 2.41, we claimed the following. Let $k \geq 2$ and $\ell \geq 1$. Let $h : A^* \to B^*$ be a $\ell$-uniform morphism. If w is a $k$-automatic word, then $h(w)$ is also $k$-automatic.

PROOF OF PROPOSITION 2.41.– For all $a \in A$, the set $\mathrm{fiber}_a(w)$ is $k$-definable by a formula $\chi_{w,a}(n)$. Then, for all $b \in B$, $\mathrm{fiber}_b(h(w))$ is $k$-definable by the formula

$$\chi_{h(w),b}(n) = (\exists m) \bigvee_{0 \leq j < \ell} \left[ (n = \ell m + j) \wedge \bigvee_a (\chi_{w,a}(m) \wedge \rho_{a,b}(j)) \right]$$

where $j$ belongs to the unary relation $\rho_{a,b}$, $a \in A$, $b \in B$, if and only if the $j$th symbol in $h(a)$ is equal to $b$. Such a relation can be finitely encoded. As an example, assume that $h(\mathsf{a}) = \mathsf{bccb}$, then $\rho_{\mathsf{a,b}}(j) \equiv (j = 0) \vee (j = 3)$ and $\rho_{\mathsf{a,c}}(j) \equiv (j = 1) \vee (j = 2)$. The meaning of the formula is the following: we divide $h(w)$ into factors of length $\ell$, and the $n$th symbol occurs in position $j$ in the $m$th block with $m, j \geq 0$. The second clause in the conjunction means that there exists some $a \in A$ such that the $m$th symbol in w is $a$. Hence, the $m$th block in $h(w)$ is $h(a)$, and we look at the $j$th symbol occurring in $h(a)$. ∎

### 3.5.2. Overlap-freeness

As an introduction, we explain how some properties about 2-automatic words, like the Thue–Morse word, can be explored (and automatically proved or disproved). In theorem 3.63, we stated that the Thue–Morse word is overlap free.

PROOF OF THEOREM 3.63 (UP TO IMPLEMENTATION).– The Thue–Morse word $\mathsf{t} = t_0 t_1 t_2 \cdots$ is 2-automatic. Hence, from corollary 3.37, there exists a formula $\chi_{t,0}(n)$ in $\langle \mathbb{N}, +, V_2 \rangle$ that holds if and only if $t_n = 0$ (i.e. a formula describing $\text{fiber}_0(t)$). The predicate

$$\text{eq}_t(i,j) \equiv (\chi_{t,0}(i) \wedge \chi_{t,0}(j)) \vee (\neg \chi_{t,0}(i) \wedge \neg \chi_{t,0}(j)) \qquad [3.5]$$

holds whenever $t_i = t_j$. Consider the sentence

$$(\exists i \geq 0)(\exists \ell \geq 1)(\forall j)(j \leq \ell \to \text{eq}_t(i+j, i+\ell+j)) \, .$$

It holds if there exists an overlap of length $2\ell + 1$ occurring in position $i$, $t_i \cdots t_{i+\ell-1} = t_{i+\ell} \cdots t_{i+2\ell-1}$ and $t_{i+\ell} = t_{i+2\ell}$. We can apply corollary 3.38 to decide whether or not this sentence is true. Of course, this requires some implementation work to really get a proof and know that t has no overlap.    ∎

EXERCISE 3.5.2.– Write a formula that holds if and only if a given $k$-automatic word contains a square (respectively, a cube, a $n$th power for a given constant $n$).

EXERCISE 3.5.3.– Write a formula that holds if and only if a given $k$-automatic word avoids squares $uu$ of length at least $|u| > K$ (for some fixed constant $K$).

EXERCISE 3.5.4.– [CHA 12] Given any $k$-automatic word x, decide whether or not x contains powers with arbitrarily large exponent.

### 3.5.3. *Abelian unbordered factors*

Let us consider another example where first-order logic can be useful. The set of lengths of abelian unbordered factors occurring in the Thue–Morse word t can be enumerated [GOČ 13a]. A key argument is that t is made up of abelian equivalent blocks 01 and 10 of length 2.

DEFINITION 3.46.– A word $u \in A^*$ is *abelian bordered* if there exists $v, v', x, y \in A^+$ such that $u = vx = yv'$ and $\Psi(v) = \Psi(v')$ where $\Psi$ is the usual Parikh map (or $v \sim_{\text{ab}} v'$ with the notation of section 3.6.1, Volume 1). In that case, $v$ is an *abelian border* of $u$. Otherwise, $u$ is said to be *abelian unbordered*. As an example, ababba is abelian bordered. The prefix ab and the suffix ba are abelian equivalent. Note that the same holds for the prefix abab and the suffix abba.

REMARK 3.47.– Note that if a word $u$ is abelian bordered, then it has an abelian border of length at most $\lfloor |u|/2 \rfloor$. Indeed, if $u = xyx'$, then $\Psi(xy) = \Psi(yx')$ implies that $\Psi(x) = \Psi(x')$.

THEOREM 3.48.– Let

$$L = 0^* \{ \mathrm{rep}_2(n) \mid \text{t has an abelian unbordered factor of length } n \} .$$

Then, $L$ is accepted by the DFA depicted in Figure 3.7, where all states are accepting except the four gray ones. Equivalently, the characteristic sequence of the set of lengths of abelian unbordered factors occurring in the Thue–Morse word is 2-automatic.

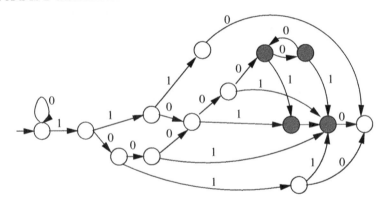

**Figure 3.7.** *A finite automaton for abelian unbordered factors in the Thue–Morse word*

PROOF (UP TO IMPLEMENTATION).– We will define the formula $\mathrm{unbordered}_t(n)$ in $\langle \mathbb{N}, +, V_2 \rangle$ which is satisfied if and only if an abelian unbordered factor of length $n$ occurs in the Thue–Morse word t.

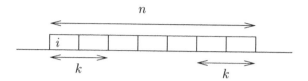

**Figure 3.8.** *Abelian border for $i, n, k$ even*

We define a ternary predicate $\text{border}_t(i, n, k)$ that is true if and only if the Thue–Morse word has an abelian bordered factor of length $n$ occurring at $i$ with a border of length $k$. As observed in remark 3.47, the fact that we choose $k \leq n/2$ will be treated in the final formula. Since the Thue–Morse word t is a concatenation of 01 and 10, we will only discuss the parity of the position $i$, the length $n$ of the factor and the length $k$ of the border. We make use of the equality modulo 2 as defined in remark 3.20. The predicate $\text{border}_t(i, n, k)$ is defined by the disjunction of the eight following terms. The first is $i \equiv_2 0 \wedge n \equiv_2 0 \wedge k \equiv_2 0$ and corresponds to the situation depicted in Figure 3.8. Recall that any two blocks of length 2 occurring in even positions are either 01 or 10, and are thus abelian equivalent. We will use the formula $\text{eq}_t(i, j)$ introduced in the proof of theorem 3.40 that holds if and only if $t_i = t_j$. Next, we have

$$i \equiv_2 0 \wedge n \equiv_2 0 \wedge k \equiv_2 1 \wedge \text{eq}_t(i + k - 1, i + n - k)$$

corresponding to the situation depicted in Figure 3.9. We simply have to check if the symbols in the $\star$ positions match. We enumerate the remaining terms (the arguments about parity are always of the same kind)

$$i \equiv_2 0 \wedge n \equiv_2 1 \wedge k \equiv_2 0 \wedge \neg\text{eq}_t(i + n - k, i + n - 1)$$

$$i \equiv_2 0 \wedge n \equiv_2 1 \wedge k \equiv_2 1 \wedge \text{eq}_t(i + k - 1, i + n - 1)$$

$$i \equiv_2 1 \wedge n \equiv_2 0 \wedge k \equiv_2 1 \wedge \text{eq}_t(i, i + n - 1)$$

$$i \equiv_2 1 \wedge n \equiv_2 1 \wedge k \equiv_2 0 \wedge \neg\text{eq}_t(i, i + k - 1)$$

$$i \equiv_2 1 \wedge n \equiv_2 1 \wedge k \equiv_2 1 \wedge \text{eq}_t(i, i + n - k).$$

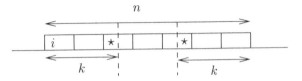

**Figure 3.9.** *Abelian border for $i, n$ even and $k$ odd*

The last term is the following one, and the corresponding situation is depicted in Figure 3.10

$$i \equiv_2 1 \wedge n \equiv_2 0 \wedge k \equiv_2 0 \wedge$$
$$((\mathsf{eq_t}(i, i + n - k) \wedge \mathsf{eq_t}(i + k - 1, i + n - 1))$$
$$\vee((\mathsf{eq_t}(i, i + n - 1) \wedge \mathsf{eq_t}(i + k - 1, i + n - k))) \, .$$

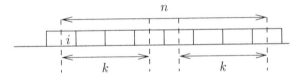

**Figure 3.10.** *Abelian border for $n, k$ even and $i$ odd*

Now, the Thue–Morse word has an abelian unbordered factor of length $n$ if and only if the following formula holds

$$\mathsf{unbordered_t}(n) \equiv (\exists i)(\forall k)(k \geq 1 \wedge 2k \leq n) \to \neg\mathsf{border_t}(i, n, k).$$

$$[3.6]$$

Since we have a formula in $\langle \mathbb{N}, +, V_2 \rangle$, we can again apply Büchi's theorem.  ∎

REMARK 3.49.– To give the readers an idea about the size of the automata that we can get, the largest intermediate step to get finally the DFA depicted in Figure 3.7 is an automaton with 215 states.

### 3.5.4. *Periodicity*

Let us now consider an emblematic decision problem. Given any $k$-automatic word $\mathbf{x} = g(f^\omega(a))$, decide whether $\mathbf{x}$ is ultimately periodic or not. Due to proposition 1.37, this problem can also be formulated as follows: given any DFA over the alphabet $[\![0, k - 1]\!]$ accepting some language $L$, decide whether or not $\mathrm{val}_k(L)$ is an ultimately periodic set of integers.

In view of Cobham's theorem (theorem 1.45), this is a problem of genuine interest. Only the ultimately periodic sets are recognizable in all integer bases. Historically, this problem has received several answers. Honkala gave a decision procedure [HON 86] precisely for $k$-automatic sets. The idea is that the number of states of the given automaton provides an upper bound for the admissible period and preperiod. Hence, we have to make comparison with elements from a finite set. Since we have Büchi's theorem at hand, therefore we can test ultimate periodicity using the sentence

$$(\exists p)(\exists N)(\forall i \geq N)\mathrm{eq}_{\mathbf{x}}(i, i + p)$$

where $\mathrm{eq}_{\mathbf{x}}$ is defined as in [3.5]. As explained in remark 3.36, this approach can lead to automata having a number of states that exponentially increases (each quantifier leads to a determinization procedure). Several authors have proposed alternatives [MAR 13, LER 05].

In the introduction of [BEL 09], about this decision problem, many pointers to the literature are given (but in a more general setting than the one of $k$-automatic sequences). In particular, we have seen that the proof of Büchi's theorem relies heavily on the fact that the graph of addition $(x, y) \mapsto x + y$ is recognized by finite automata (for both integer base systems and Pisot numeration systems). Nevertheless, out of the Pisot setting, a decision procedure may exist even if the graph of addition is not recognized by finite automata. Given a pure morphic word (where the iterated morphism is not necessarily uniform), we can decide whether or not this word is ultimately periodic [HAR 86, PAN 86]. Durand proved that the problem turns out to be decidable for arbitrary morphic words [DUR 12, DUR 13a]. The proofs do not use methods from formal logic but rely on classical tools such as return words (see section 3.2, Volume 1). Roughly speaking, we can try to give computable upper bounds on the possible period and preperiod that the morphic word might have, and thus we have to make a finite number of tests to decide about ultimate periodicity. Mitrofanov obtained the same result independently [MIT 11] by analyzing Rauzy graphs.

### 3.5.5. *Factors*

With such a formal tool for automatic verification, in a series of papers with several coauthors Goč and Shallit investigated properties of $k$-automatic sequences that can be expressed in $\langle \mathbb{N}, +, V_k \rangle$. See [GOČ 12, GOČ 13b].

Let $x, y$ be two $k$-automatic words over the same alphabet. We define

$$\mathsf{same}_{x,y}(i, j, \ell) \equiv (\forall t)(t < \ell \to x(i + t) = y(j + t))$$

to express the fact that the factor of $x$ of length $\ell$ occurring in position $i$ is equal to the factor of $y$ of the same length occurring in position $j$. Similarly to the formula $\mathsf{eq}_x(i, j)$ introduced in the proof of theorem 3.40, we note that $x(i) = y(j)$ is definable in $\langle \mathbb{N}, +, V_k \rangle$ by

$$\bigvee_{a \in A} (\chi_{x,a}(i) \wedge \chi_{y,a}(j))$$

where the predicate $\chi_{x,a}(n)$ given by corollary 3.37 holds if and only if $x(n) = a$.

PROPOSITION 3.50.– Given two $k$-automatic sequences x and y, we can decide whether or not

 – $\mathrm{Fac}\,\mathrm{x} \supseteq \mathrm{Fac}\,\mathrm{y}$;

 – $\mathrm{Fac}\,\mathrm{x} = \mathrm{Fac}\,\mathrm{y}$;

 – x is recurrent.

The first part of the statement was considered in [FAG 97].

PROOF.– The sentence $(\forall i)(\forall n)(\exists j)\mathsf{same}_{\mathsf{x},\mathsf{y}}(i,j,n)$ holds if and only if $\mathrm{Fac}\,\mathrm{x} \supseteq \mathrm{Fac}\,\mathrm{y}$.

Recall that an infinite word x is recurrent if and only if every factor of x occurs at least twice (see exercise 3.2.3, Volume 1). Therefore, we have to check the following sentence

$$(\forall i)(\forall n)(\exists j)(\neg(i = j) \wedge \mathsf{same}_{\mathsf{x},\mathsf{x}}(i,j,n))\,.$$

∎

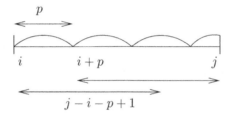

**Figure 3.11.** *A finite word with period $p$*

A finite word $w = w_0 \cdots w_r$ has *period*[19] $p > 0$ if $w_i = w_{i+p}$ for all $i \in [\![0, r - p]\!]$. Let x be a $k$-automatic word. Hence, the formula

$$\mathsf{per}_{\mathsf{x}}(p,i,j) \equiv p \geq 1 \wedge p \leq j - i \wedge \mathsf{same}_{\mathsf{x},\mathsf{x}}(i, i + p, j - i - p + 1)$$

holds if the finite word $\mathsf{x}(i) \cdots \mathsf{x}(j)$ has period $p$ as depicted in Figure 3.11. To express that $p$ is the least period of $\mathsf{x}(i) \cdots \mathsf{x}(j)$, write $\mathsf{per}_{\mathsf{x}}(p,i,j) \wedge (\forall n)(n < p \to \neg\mathsf{per}_{\mathsf{x}}(n,i,j))$.

---

19 Have a look at theorem 1.77, Volume 1.

PROPOSITION 3.51.– Let $k \geq 2$. Given a $k$-automatic sequence x, the characteristic sequence of the set of integers that are the least period of a factor occurring in x is $k$-automatic.

PROOF.– The set $\{p \mid \mathsf{per}_{\mathsf{x}}(p, i, j) \wedge (\forall n)(n < p \rightarrow \neg \mathsf{per}_{\mathsf{x}}(n, i, j))\}$ is $k$-definable, and thus $k$-recognizable.    ∎

Based on Büchi's theorem, Charlier, Rampersad and Shallit considered enumeration problems [CHA 12]. Let x be a $k$-automatic word. The formula

$$\mathsf{firstOcc}(i, n) \equiv (\forall j)(j < i \rightarrow \neg \mathsf{same}_{\mathsf{x},\mathsf{x}}(i, j, n))$$

holds if and only if the factor $x_i \cdots x_{i+n-1}$ of length $n$ occurs for the first time in position $i$. From Büchi's theorem, we know that the set $\{(i, n) \mid n_k \models \mathsf{firstOcc}(i, n)\}$ is $k$-recognizable. The authors, therefore, used this result to show that the sequence $(\mathsf{p}_{\mathsf{x}}(n))_{n \geq 0}$ is $(\mathbb{N}, k)$-regular. See section 2.8, Volume 1. Many examples and applications are presented in [CHA 12].

EXERCISE 3.5.5.– Consider the 2-block coding of the Thue–Morse word. It is the fixed point starting with 1 of the morphism $1 \mapsto 13$, $2 \mapsto 20$, $3 \mapsto 21$ and $0 \mapsto 12$. This word b starts with $132120132012132 \cdots$. Write a formula to prove or disprove that $g(\mathbf{b}) = (12)^{\omega}$ and $h(\mathbf{b}) = (30)^{\omega}$ where $g(0) = g(3) = \varepsilon$, $g(1) = 1$ and $g(2) = 2$, $h(1) = h(2) = \varepsilon$, $h(0) = 0$ and $h(3) = 3$.

### 3.5.6. Applications to Pisot numeration systems

We start with an example. Consider the morphism $f : \{q, r, s, t\}^* \rightarrow \{q, r, s, t\}^*$ given by

$$f : q \mapsto qr, r \mapsto s, s \mapsto st, t \mapsto q$$

and the coding $g : q, t \mapsto 1, s, r \mapsto 0$. The first few symbols of $\mathbf{w} = g(f^\omega(q))$ are $1000101101110\cdots$. Can we use the formalism of first-order logic to derive some properties (like appearance of a repetition or ultimate periodicity) about such a morphic word $\mathbf{w}$?

With the morphism $f$ is canonically associated a DFA $\mathcal{A}_{f,q}$ where all states are final. See, for instance, the proof of proposition 2.29. The transitions of $\mathcal{A}_{f,q}$ are exactly those depicted in Figure 2.14. We have seen in example 2.53 that, if all states of this automaton are assumed to be final, then the accepted language is exactly $0^* \mathrm{rep}_U(\mathbb{N})$ where $U$ is the Zeckendorf numeration system. Again from the proof of proposition 2.29, feeding $\mathcal{A}_{f,q}$ from the initial state $q$ with the normal $U$-representation $\mathrm{rep}_U(n)$, which does not start with $0$, we reach the state $(f^\omega(q))(n)$. This means that, for all $p \in \{q, r, s, t\}$, the set $\mathrm{fiber}_{f^\omega(q)}(p)$ is $U$-recognizable using the DFA $\mathcal{A}_{f,q}$ where the unique final state is set to $p$.

Recall that the Zeckendorf numeration system is a Pisot system associated with the golden ratio. Therefore, from theorem 3.45, for all $p \in \{q, r, s, t\}$, the sets $\mathrm{fiber}_{f^\omega(q)}(p)$ and thus $\mathrm{fiber}_\mathbf{w}(g(p))$ are $U$-definable. We can, therefore, apply the same tests as those described in the previous sections.

REMARK 3.52.– The word $\mathbf{w} = 1000101101110\cdots$ is the characteristic sequence of the set of integers whose Zeckendorf expansion contains an even number of 1's.

| 1 | 1 | 5 | 1000 | 9 | 10001 |
|---|---|---|------|---|-------|
| 2 | 10 | 6 | 1001 | 10 | 10010 |
| 3 | 100 | 7 | 1010 | 11 | 10100 |
| 4 | 101 | 8 | 10000 | 12 | 10101 |

**Table 3.1.** *Zeckendorf expansions of the first 12 integers*

What we have observed in this example can be formalized as follows.

PROPOSITION 3.53.– Let $U$ be a Pisot numeration system. Let $f : A^* \to A^*$ be a morphism prolongable on $a \in A$. Let $g : A^* \to B^*$ be a coding. Let $\mathbf{w} = g(f^\omega(a)) = w_0 w_1 w_2 \cdots$. If the language accepted by $\mathcal{A}_{f,a}$ (where all states are final) is equal to $0^* \operatorname{rep}_U(\mathbb{N})$, then the set $\operatorname{fiber}_{\mathbf{w}}(b) = \{n \mid w_n = b\}$ is $U$-definable for all $b \in B$.

EXAMPLE    3.54.– Consider    the    Tribonacci    word $\mathcal{T} = $ abacabaabacab$\cdots$ fixed point of the morphism $\mu : $ a $\mapsto$ ab, b $\mapsto$ ac, c $\mapsto$ a. The associated automaton $\mathcal{A}_{\mu,a}$ is given in Figure 2.18. The positional numeration system $T = (T_n)_{n \geq 0}$ defined by $T_0 = 1$, $T_1 = 2$, $T_2 = 4$ and $T_{n+3} = T_{n+2} + T_{n+1} + T_n$, for all $n \geq 0$, is a Pisot numeration system. Example 2.66 shows that the sets $\operatorname{fiber}_{\mathcal{T}}(a)$ for $a \in \{$a, b, c$\}$ are $T$-recognizable, and therefore $T$-definable by a formula $\chi_{T,a}$.

Another way to recover the same result is to observe that the DFA $\mathcal{A}_{\mu,a}$ recognizes exactly $0^* \operatorname{rep}_T(\mathbb{N})$ and apply the previous proposition.

The Tribonacci word is an Arnoux–Rauzy word (see example 3.33, Volume 1). In particular, for every $n$, $\mathcal{T}$ has a exactly one right special factor of length $n$ with three extensions and every other factor of length $n$ has exactly one extension. This implies that its factor complexity satisfies $\mathsf{p}_{\mathcal{T}}(n) = 2n + 1$ for all $n$. Such a statement can be expressed in $\langle \mathbb{N}, +, V_T \rangle$ by the formula

$$(\forall n)(\exists i)(\exists j)(\exists k)(\mathsf{same}_{\mathcal{T},\mathcal{T}}(i,j,n) \wedge \mathsf{same}_{\mathcal{T},\mathcal{T}}(j,k,n)$$

$$\wedge \chi_{\mathcal{T},\mathsf{a}}(i+n) \wedge \chi_{\mathcal{T},\mathsf{b}}(j+n) \wedge \chi_{\mathcal{T},\mathsf{c}}(k+n) \wedge$$

$$(\forall p)(\forall q)((\neg \mathsf{same}_{\mathcal{T},\mathcal{T}}(i,p,n) \wedge \mathsf{same}_{\mathcal{T},\mathcal{T}}(p,q,n))$$

$$\to \mathsf{eq}_T(p+n, q+n)).$$

This formula is interpreted as follows. There exists three positions $i, j, k$ where the same factor $u$ of length $n$ occurs, but this factor is followed either by a, b or c. So, the first two lines of the formula are expressing the existence of a special factor of length $n$ with extensions. Then for every factor $v$ of length $n$ which is distinct from $u$ (the one occurring in position $i$), $v$ is always followed by the same symbol: if $v$ occurs in two positions $p$ and $q$, then $\mathcal{T}(p + n) = \mathcal{T}(q + n)$.

EXERCISE 3.5.6.– Write a formula expressing that the Fibonacci word has exactly one right special factor of each length.

To conclude the proof of corollary 2.76, Volume 1, we can use the formal methods developed in this section.

EXERCISE 3.5.7.– Write a formula expressing that the Fibonacci (respectively, Tribonacci) word is not ultimately periodic.

## 3.6. Bibliographic notes and comments

Again, [BRU 94] is a very useful reference. It contains many pointers to the literature, extended bibliographic notes and details about the proof of Büchi's theorem. It also presents a proof of Cobham–Semenov theorem (see theorem 3.42) for two integer bases and subsets of $\mathbb{N}^n$. About Büchi's theorem, see see [HOD 83]. About definability and decidability issues, see also the survey [BÈS 01] where in particular the *Skolem arithmetic* $\langle \mathbb{N}, +, \times \rangle$ is introduced.

Regarding **complexity issues** raised by Büchi's theorem when replacing a formula with a DFA, we should consider [KLA 08] where an upper bound on the number of states of the minimal automaton derived from a formula in Presburger arithmetic is given. Naturally, this bound depends on the length of the formula and the quantifiers occurring in it. Interestingly, F. Klaedtke compared the two approaches discussed in this chapter: elimination of quantifiers (theorem 3.24) and automata derived from formula (see remark 3.39). Another paper on the same topic is [OPP 78]. It is also worth mentioning the paper [LER 05] where a polynomial decision procedure is provided to test if a $k$-definable set in $\mathbb{N}^n$ is definable in Presburger arithmetic.

Much more can be said about the conversion between automata and formulas. In [BOI 04], the authors applied a projection method to efficiently construct automata from quantified formulas.

We can also be interested in recognizable or definable **sets of real numbers**. Using integer base-$b$ expansions and with suitable conventions to take care of the decimal point, a real number is represented by an infinite word over $[\![0, b-1]\!]$. A set of real numbers is $b$-recognizable if the language of its base-$b$ expansions is recognized by some Büchi automaton. In this context, a variant of Büchi's theorem can be stated. Hence, infinite paths in an automaton are coded by real numbers. See [BOI 98]. For extensions to $\beta$-numeration, see [CHA 13]. We observed that extending Büchi's theorem from integer base systems to Pisot numeration systems leads to some technical refinements about the admissible codings (see the proof of theorem 3.45). The same kind of discussion appears between integer base and $\beta$-numeration systems.

For Cobham's theorem on the base dependence and more generally regarding the **Cobham–Semenov theorem**, the logical formalism turns out to be useful. See, for instance, [MIC 96]. In [BÈS 00], the result is extended to Pisot numeration systems where the associated real numbers are multiplicatively independent. An intermediate step is to consider one Pisot system and one integer base system [POI 97].

For Cobham's theorem in the context of sets of real numbers recognizable in integer bases, see [BOI 08, BOI 09a, BOI 09b].

We have seen that the first-order theory of $\langle \mathbb{N}, +, V_k \rangle$ is decidable. Regarding **undecidable theories**, see [VIL 92] and the generalization [BÈS 97]. The first-order theory of $\langle \mathbb{N}, +, V_k, V_\ell \rangle$ is undecidable whenever $k, \ell \geq 2$ are multiplicatively independent. Recall that it is also a well-known result that $\langle \mathbb{N}, +, \times \rangle$ has an undecidable first-order theory. J. Barkley Rosser proved this result in 1937, J. obinson proved a similar result for the set of rational numbers in her thesis in 1948. In that direction, the readers may also consult the book *Undecidable theories* (1953) of A. Tarski in collaboration with A. Mostowski and J. Robinson for a rather general method for proving undecidability of a wide class of theories.

Even though the expressiveness of Büchi's theorem can be considered as satisfactory (we have seen plenty of examples), note that some properties cannot be checked with first-order logic. We have seen that expressing the fact that a $k$-automatic word x contains a square in position $i$ can be handled by the following formula in $\langle \mathbb{N}, +, V_k \rangle$

$$\varphi(i) \equiv (\exists n)(n > 0 \wedge \mathsf{same}_{\mathsf{x},\mathsf{x}}(i, i + n, n)) .$$

L. Schaeffer showed that the set of occurrences of abelian squares (see exercise 3.6.2, Volume 1) in the paperfolding word[20] is not 2-recognizable, even though the paperfolding word is 2-automatic. Hence, such a property cannot be captured by a formula in $\langle \mathbb{N}, +, V_2 \rangle$.

The link between logic and regular languages is stronger than what has been presented in this chapter. We know that regular languages are recognized by finite automata and described by regular expressions. Since the work of Büchi [BÜC 60], it is known that regular languages also can be defined using monadic second-order logic (Büchi–Elgot

---

20 See exercise 2.7.18 and example 3.36, both in Volume 1.

theorem). The adjective *second-order* means that we can quantify over relations. The extra qualification *monadic* means that we can quantify over unary relations only, i.e. over sets of variables. Variables represent positions within a word, and we have special predicates such as $a(n)$ or $b(n)$ to express that a given symbol a or b appears in a specific position $n \geq 0$. In this setting, we can quantify not only over variables, but also over the sets of variables. A finite word $w$ gives an iterpretation of a sentence, where the variables belong to $[0, |w| - 1]$ and sets belong to $2^{[0,|w|-1]}$. Also, there is a special predicate $\texttt{last}(n)$ that holds if $n$ is equal to $|w| - 1$. For example, the (first-order) formula

$$(\forall n)(\forall m)((b(n) \wedge n < m) \to b(m))$$

is interpreted as follows: for every position $n$ where the letter b occurs, every other symbol occurring to its right, if any, should also be equal to b. The language a*b* is defined by this formula. For example, using second-order quantification, considers the language of words of even length. We can define two sets of positions: the set $X_e$ of even positions and the set $X_o$ of odd positions. Consider the sentence

$$(\exists X_e)(\exists X_o)(\forall n) \left( ((n \in X_e) \leftrightarrow (n \notin X_o)) \wedge ((n = 0) \to n \in X_e) \wedge \right.$$

$$((\forall m)((n \in X_e) \wedge (m = n + 1)) \to m \in X_o) \wedge$$

$$((\forall m)((n \in X_o) \wedge (m = n + 1)) \to m \in X_e) \wedge$$

$$\left. (\texttt{last}(n) \to (n \in X_o)) \right).$$

More can be said: extensions to infinite words and applications to dedicable theories. See, for instance, [STR 94].

# List of Sequences

The On-Line Encyclopedia of Integer Sequences (OEIS) http://oeis.org/ is a useful tool. Anyone seriously interested in combinatorics of sequences should know about it. We can get relevant information, such as the generating function, on many sequences we might encounter. We have collected the sequences of this book that are recorded in the OEIS. The list is presented in chronological order of appearance within this book. We refer to the name of the sequence found in the encyclopedia.

From example 1.39, Volume 1, on Pascal's triangle – #A047999 *Pascal's triangle mod 2*,

$b_2 = 1111011111100011100111010101111111111100000001\cdots$ ;

#A083093 *Pascal's triangle mod 3*,

$b_3 = 1\ 11\ 121\ 1001\ 11011\ 121121\ 1002001\ 11022011\ 121212121\cdots$ ;

#A095142 *Pascal's triangle mod 7*,

$b_7 = 1111211331146411533511616161100000011110000011\cdots$ .

From example 1.41, Volume 1, – #A030302 *Write n in base 2 and juxtapose*, essentially the same as #A030190,

$$\mathbf{c}_2 = 11011100101110111110001001101 \cdots .$$

From example 1.54, Volume 1, – #A003849 *The infinite Fibonacci word*,

$$\mathbf{f} = 010010100100101001010010010010010 \cdots .$$

From example 2.1, Volume 1, – #A010060 *Thue–Morse sequence*,

$$\mathbf{t} = 0110100110010110100101100110100110010110011010 01 \cdots .$$

From example 2.5, Volume 1, – #A010052 *Characteristic sequence of squares*,

$$1100100001000000100000000100000000000100 \cdots .$$

From section 2.2.2, Volume 1, – #A092782 *The Tribonacci word*,

$$1213121121312121312112131213121121312121312112131211213121 \cdots$$

From example 2.34, Volume 1, about $(\mathsf{p_t}(n))_{n \geq 0}$ – #A005942 *The factor complexity of the Thue–Morse word*,

$$1, 2, 4, 6, 10, 12, 16, 20, 22, 24, 28, 32, 36, 40, 42, 44, 46, 48, 52, 56, \ldots$$

From exercise 2.7.18, Volume 1, – #A014577 *The regular paper folding (or dragon curve) sequence*,

$$110110011100100 \cdots$$

From example 2.94, Volume 1, – #A000120 *1's-counting sequence: number of 1's in binary expansion of n*,

$$0, 1, 1, 2, 1, 2, 2, 3, 1, 2, 2, 3, \ldots .$$

From example 2.95, Volume 1; see also example 3.36, Volume 1, – #A003602 *Kimberling's paraphrases,*

$$1, 1, 2, 1, 3, 2, 4, 1, 5, 3, 6, \ldots .$$

From example 2.96, Volume 1, – #A006257 *Josephus problem,*

$$1, 1, 3, 1, 3, 5, 7, 1, 3, 5, 7, 9, 11, 13, 15, 1, 3, \ldots .$$

From section 3.3, Volume 1, – #A000002 *Kolakoski sequence,*

$$k = 22112122122112112212112122112112122122112122112 \cdots$$

From example 3.32, Volume 1, – #A007814 *Exponent of highest power of 2 dividing n, a.k.a. the binary carry sequence, the ruler sequence, or the 2-adic valuation of n,*

$$r = 0102010301020104010201030102010 5 \cdots .$$

As one example of application of proposition 3.58, Volume 1, – #A214212 *Number of special factors of length n in the Thue–Morse sequence,*

$$2, 2, 4, 2, 4, 4, 2, 2, 4, 4, 4, 4, 2, 2, 2, 2, \ldots$$

From corollary 3.64, Volume 1, – #A007413 *A squarefree (or Thue–Morse) ternary sequence,*

$$1231321232131 \cdots$$

From theorem 3.91, Volume 1, – #A096268 *Period-doubling sequence,*

$$0100010101000100010001010100010101000101010 00100010001 \cdots ,$$

and thus – #A216195 *Abelian complexity function of the period-doubling sequence,*

$$1, 2, 2, 3, 2, 3, 3, 3, 2, 3, 3, 4, 3, 4, 3, 3, 2, 3, 3, 4, 3, 4, 4, 4, 3, 4, 4, 4, 3, 4, 3, 3, \ldots.$$

From section 3.7, Volume 1, – #A191818 *A sum-square avoiding sequence,*

$$0314301103434303101101103143034343034343034303143011031011011 \cdots$$

From section 1.7, #A209229 *Characteristic function of powers of 2,*

$$0110100010000000100000000000000010000000000000000000000000000 \cdots.$$

From example 1.59 – #A000124 *Central polygonal numbers (the Lazy Caterer's sequence),*

$$1, 2, 4, 7, 11, 16, 22, 29, 37, 46, 56, 67, 79, 92, 106, 121, 137, \ldots.$$

From example 2.11 – #A000213 *Tribonacci numbers,*

$$1, 1, 1, 3, 5, 9, 17, 31, 57, 105, 193, 355, 653, 1201, 2209, 4063, 7473, 13745, \ldots.$$

and similarly – #A000288 *Tetranacci numbers,*

$$1, 1, 1, 1, 4, 7, 13, 25, 49, 94, 181, 349, 673, 1297, 2500, 4819, 9289, 17905, \ldots.$$

From section 2.3 – #A000045 *Fibonacci numbers,*

$$1, 2, 3, 5, 8, 13, 21, 34, 55, 89, 144, 233, 377, 610, 987, 1597, 2584, 4181, \ldots.$$

From example 2.47 – #A000290 *The squares,*

$$1, 4, 9, 16, 25, 36, 49, 64, 81, 100, 121, 144, 169, 196, 225, 256, \ldots.$$

From exercise 2.3.2 – #A204418 *Periodic sequence,*

$(1, 0, 1)^\omega$.

From example 2.58 – #A000032 *Lucas numbers,*

$1, 3, 4, 7, 11, 18, 29, 47, 76, 123, 199, 322, 521, 843, 1364, 2207, 3571, 5778, \ldots$.

From remark 3.52 – #A095111 *One minus the parity of 1-fibits[1] in Zeckendorf expansion,*

$1, 0, 0, 0, 1, 0, 1, 1, 0, 1, 1, 1, 0, 0, 1, 1, 1, 0, 1, 0, 0, 0, 1, 1, 1, 0, 1, 0, 0, \ldots$.

---

[1] Stands for Fibonacci digits.

# Bibliography

[ABE 04] ABERKANE A., CURRIE J.D., RAMPERSAD N., "The number of ternary words avoiding abelian cubes grows exponentially", *Journal of Integer Sequences,* vol., 7, no. 2, Article 04.2.7, 2004.

[ADA 02] ADAMCZEWSKI B., "Codages de rotations et phénomènes d'autosimilarité", *Journal de Théorie des Nombres de Bordeaux,* vol. 14, no. 2, pp. 351–386, 2002.

[ADA 03] ADAMCZEWSKI B., "Balances for fixed points of primitive substitutions", *Theoretical Computer Science,* vol. 307, no. 1, pp. 47–75, 2003.

[ADA 04a] ADAMCZEWSKI B., BUGEAUD Y., LUCA F., "Sur la complexité des nombres algébriques", *Comptes Rendus de l'Académie des Sciences Paris,* vol. 339, pp. 11–14, 2004.

[ADA 04b] ADAMCZEWSKI B., "Symbolic discrepancy and self-similar dynamics", *Annales de l'Institut Fourier (Grenoble),* vol. 54, no. 7, pp. 2201–2234, 2004.

[ADA 07] ADAMCZEWSKI B., BUGEAUD Y., "On the complexity of algebraic numbers. I. Expansions in integer bases", *Annals of Mathematics,* vol. 165, pp. 547–565, 2007.

[ADA 12] ADAMCZEWSKI B., BELL J.P., "On vanishing coefficients of algebraic power series over fields of positive characteristic", *Inventiones Mathematicae,* vol. 187, no. 2, pp. 343–393, 2012.

[AKI 99] AKIYAMA S., "Self affine tiling and Pisot numeration system", in KANEMITSU S., GYÖRY K., (eds.), *Number Theory and Its Applications*, Kluwer, pp. 7–17, 1999.

[AKI 07] AKIYAMA S., SHIRASAKA M., "Recursively renewable words and coding of irrational rotations", *Journal of the Mathematical Society of Japan,* vol. 59, no. 4, pp. 1199–1234, 2007.

[AKI 08] AKIYAMA S., FROUGNY C., SAKAROVITCH J., "Powers of rationals modulo 1 and rational base number systems", *Israel Journal of Mathematics,* vol. 168, pp. 53–91, 2008.

[ALA 04] ALACA Ş., WILLIAMS K.S., *Introductory algebraic number theory*, Cambridge University Press, Cambridge, 2004.

[ALE 98] ALESSANDRI P., BERTHÉ V., "Three distance theorems and combinatorics on words", *L'Enseignement Mathématique,* vol. 44, pp. 103–132, 1998.

[ALE 04] ALEXEEV B., "Minimal DFA for testing divisibility", *Journal of Computer and System Sciences*, vol. 69, no. 2, pp. 235–243, 2004.

[ALL 92a] ALLOUCHE J.-P., SHALLIT J.O., "The ring of $k$-regular sequences", *Theoretical Computer Science,* vol. 98, pp. 163–197, 1992.

[ALL 92b] ALLOUCHE J.-P., "$q$-regular sequences and other generalizations of $q$-automatic sequences", in *LATIN '92*, Lecture Notes in Computer Science, Springer-Verlag, Berlin, vol. 583, pp. 15–23, 1992.

[ALL 96] ALLOUCHE J.-P., VON HAESELER F., PEITGEN H.-O., SKORDEV G., "Linear cellular automata, finite automata and Pascal's triangle", *Discrete Applied Mathematics,* vol. 66, pp. 1–22, 1996.

[ALL 97a] ALLOUCHE J.-P., BERTHÉ V., "Triangle de Pascal, complexité et automates", *Bulletin of the Belgian Mathematical Society,* vol. 4, pp. 1–23, 1997.

[ALL 97b] ALLOUCHE J.-P., VON HAESELER F., LANGE E., PETERSEN A., SKORDEV G., "Linear cellular automata and automatic sequences", *Parallel Computing,* vol. 23, no. 11, pp. 1577–1592, 1997.

[ALL 97c] ALLOUCHE J.-P., VON HAESELER F., PEITGEN H.-O., PETERSEN A., SKORDEV G., "Automaticity of double sequences generated by one-dimensional linear cellular automata", *Theoretical Computer Science*, vol. 188, pp. 195–209, 1997.

[ALL 98] ALLAUZEN C., "Une caractérisation simple des nombres de Sturm", *Journal de Théorie des Nombres de Bordeaux*, vol. 10, pp. 237–241, 1998.

[ALL 99a] ALLOUCHE J.-P., SHALLIT J.O., "The ubiquitous Prouhet-Thue-Morse sequence", in DING C., HELLESETH T., NIEDERREITER H. (eds.), *Sequences and Their Applications, Proceedings of SETA '98*, Springer-Verlag, pp. 1–16, 1999.

[ALL 99b] ALLOUCHE J.-P., 1999, Research Seminar, University of Brussels.

[ALL 03a] ALLOUCHE J.-P., SHALLIT J., *Automatic Sequences: Theory, Applications, Generalizations*, Cambridge University Press, Cambridge, 2003.

[ALL 03b] ALLOUCHE J.-P., SHALLIT J., "The ring of $k$-regular sequences. II", *Theoretical Computer Science*, vol. 307, no. 1, pp. 3–29, 2003.

[ALL 08] ALLOUCHE J.-P., MENDÈS FRANCE M., "Euler, Pisot, Prouhet-Thue-Morse, Wallis and the duplication of sines", *Monatshefte für Mathematik*, vol. 155, nos. 3–4, pp. 301–315, 2008.

[AMB 12] AMBROŽ P., DOMBEK D., MASÁKOVÁ Z., PELANTOVÁ E., "Numbers with integer expansion in the numeration system with negative base", *Functiones et Approximatio, Commentarii Mathematici*, vol. 47, part 2, pp. 241–266, 2012.

[ANG 11] ANGRAND P.-Y., SAKAROVITCH J., "On the enumerating series of an abstract numeration system", 2011. Available at http://arxiv.org/abs/1108.5711

[ARN 91] ARNOUX P., RAUZY G., "Représentation géométrique de suites de complexité $2n + 1$", *Bulletin de la Société Mathématique de France*, vol. 119, pp. 199–215, 1991.

[ARN 01] ARNOUX P., ITO S., "Pisot substitutions and Rauzy fractals", *Bulletin of the Belgian Mathematical Society*, vol. 8, pp. 181–207, 2001.

[ARN 04a] ARNOUX P., SIEGEL A., Dynamique du nombre d'or, Actes de l'université d'été de Bordeaux, 2004.

[ARN 04b] ARNOUX P., BERTHÉ V., SIEGEL A., "Two-dimensional iterated morphisms and discrete planes", *Theoretical Computer Science*, vol. 319, nos. 1–3, 145–176, 2004.

[ARO 09] ARORA S., BARAK B., *Computational Complexity, A Modern Approach*, Cambridge University Press, Cambridge, 2009.

[AVG 03] AVGUSTINOVICH S.V., FON-DER-FLAASS D.G., FRID A.E., "Arithmetical complexity of infinite words", in *Words, Languages & Combinatorics, III (Kyoto, 2000)*, World Science Publisher, River Edge, NJ, pp. 51–62, 2003.

[AVG 06] AVGUSTINOVICH S.V., CASSAIGNE J., FRID A., "Sequences of low arithmetical complexity", *RAIRO Theoretical Informatics and Applications*, vol. 40, no. 4, pp. 569–582, 2006.

[AVG 11] AVGUSTINOVICH S.V., FRID A., KAMAE T., SALIMOV P., "Infinite permutations of lowest maximal pattern complexity", *Theoretical Computer Science*, vol. 412, no. 27, pp. 2911–2921, 2011.

[AVI 61] AVIZIENIS A., "Signed-digit number representations for fast parallel arithmetic", *IRE Transactions on Electronic Computers*, vol. 10, pp. 389–400, 1961.

[BAC 96] BACH E., SHALLIT J., *Algorithmic Number Theory*, MIT Press, 1996.

[BAL 08] BALÁŽI P., MASÁKOVÁ Z., PELANTOVÁ E., "Characterization of substitution invariant words coding exchange of three intervals", *Integers*, vol. 8, A20, 2008.

[BAL 13] BALKOVÁ A., BUCCI M., DE LUCA L., PUZYNINA S., "Infinite words with well distributed occurrences", *WORDS 2013*, Lecture Notes in Computer Science, vol. 8079, Springer-Verlag, pp. 46–57, 2013.

[BAR 87] BARBIER E., "On suppose écrite la suite naturelle des nombres; quel est le $(10^{1000})^{\text{ième}}$ chiffre écrit?", *Comptes Rendus de l'Académie des Sciences Paris*, vol. 105, pp. 795–798, 1887.

[BAR 06a] BARAT G., BERTHÉ V., LIARDET P., THUSWALDNER J., "Dynamical directions in numeration", *Annales de l'Institut Fourier (Grenoble)*, vol. 56, no. 7, pp. 1987–2092, 2006.

[BAR 06b] BARGE M., KWAPISZ J., "Geometric theory of unimodular Pisot substitutions", *American Journal of Mathematics,* vol. 128, no. 5, pp. 1219–1282, 2006.

[BEA 79] BEAN D.A., EHRENFEUCHT A., MCNULTY G., "Avoidable patterns in strings of symbols", *Pacific Journal of Mathematics,* vol. 85, pp. 261–294, 1979.

[BEC 84] BECK J., "An application of Lovász local lemma: there exists an infinite 01-sequence containing no near identical intervals", in *Finite and Infinite Sets, vol. I, II (Eger, 1981),* János Bolyai Mathematical Society, North-Holland, Amsterdam, vol. 37, pp. 103–107, 1984.

[BEL 07a] BELL J.P., "A generalization of Cobham's theorem for regular sequences", *Séminaire Lotharingien de Combinatoire,* vol. 54A, B54A, 07/2005.

[BEL 05] BELL J.P., "On the values attained by a $k$-regular sequence", *Advances in Applied Mathematics,* vol. 34, no. 3, pp. 634–643, 2005.

[BEL 07b] BELL J.P., GOH T.L., "Exponential lower bounds for the number of words of uniform length avoiding a pattern", *Information and Computation,* vol. 205, no. 9, pp. 1295–1306, 2007.

[BEL 08] BELL J.P., "Logarithmic frequency in morphic sequences", *Journal de Théorie des Nombres de Bordeaux,* vol. 20, no. 2, pp. 227–241, 2008.

[BEL 09] BELL J., CHARLIER É., FRAENKEL A.S., RIGO M., "A decision problem for ultimately periodic sets in nonstandard numeration systems", *International Journal of Algebra and Computation,* vol. 19, no. 6, pp. 809–839, 2009.

[BER 72] BERNOULLI J., "Sur une nouvelle espèce de calcul", In *Recueil pour les Astronomes,* vol. I, pp. 255–284, 1772.

[BER 77] BERTRAND A., "Développements en base de Pisot et répartition modulo 1", *Comptes rendus de l'Académie des Sciences, Séries A-B,* vol. 285, no. 6, pp. A419–A421, 1977.

[BER 89] BERTRAND-MATHIS A., "Comment écrire les nombres entiers dans une base qui n'est pas entière", *Acta Mathematica Hungarica,* vol. 54, pp. 237–241, 1989.

[BER 93] BERSTEL J., POCCHIOLA M., "A geometric proof of the enumeration formula for Sturmian words", *International Journal of Algebra and Computation*, vol. 3, pp. 349–355, 1993.

[BER 94a] BEREND D., FROUGNY C., "Computability by finite automata and Pisot bases", *Mathematical Systems Theory*, vol. 27, pp. 275–282, 1994.

[BER 94b] BERSTEL J., SÉÉBOLD P., "Morphismes de Sturm", *Bulletin of the Belgian Mathematical Society*, vol. 1, pp. 175–189, 1994.

[BER 97] BERSTEL J., BOASSON L., "The set of minimal words of a context-free language is context-free", *Journal of Computer and System Sciences*, vol. 55, no. 3, pp. 477–488, 1997.

[BER 00] BERTHÉ V., VUILLON L., "Tilings and rotations on the torus: a two-dimensional generalization of Sturmian sequences", *Discrete Mathematics*, vol. 223, pp. 27–53, 2000.

[BER 01] BERTHÉ V., "Autour du système de numération d'Ostrowski", *Bulletin of the Belgian Mathematical Society*, vol. 8, pp. 209–239, 2001.

[BER 02] BERSTEL J., VUILLON L., "Coding rotations on intervals", *Theoretical Computer Science*, vol. 281, nos. 1–2, pp. 99–107, 2002.

[BER 05a] BERSTEL J., "Growth of repetition-free words–a review", *Theoretical Computer Science*, vol. 340, no. 2, pp. 280–290, 2005.

[BER 05b] BERTHÉ V., BRLEK S., CHOQUETTE P., "Smooth words over arbitrary alphabets", *Theoretical Computer Science*, vol. 341, nos. 1–3, pp. 293–310, 2005.

[BER 05c] BERTHÉ V., SIEGEL A., "Tilings associated with beta-numeration and substitutions", *Integers*, vol. 5, no. 3, A2, 2005.

[BER 06a] BERSTEL J., BOASSON L., CARTON O., PETAZZONI B., PIN J.-E., "Operations preserving regular languages", *Theoretical Computer Science*, vol. 354, no. 3, pp. 405–420, 2006.

[BER 06b] BERTHÉ V., HOLTON C., ZAMBONI L.Q., "Initial powers of Sturmian sequences", *Acta Arithmetica*, vol. 122, no. 4, pp. 315–347, 2006.

[BER 07a] BERNAT J., MASÁKOVÁ Z., PELANTOVÁ E., "On a class of infinite words with affine factor complexity", *Theoretical Computer Science*, vol. 389, nos. 1–2, pp. 12–25, 2007.

[BER 07b]  BERSTEL J., PERRIN D., "The origins of combinatorics on words", *European Journal of Combinatorics,* vol. 28, no. 3, pp. 996–1022, 2007.

[BER 07c]  BERTHÉ V., EI H., ITO S., RAO H., "On substitution invariant Sturmian words: an application of Rauzy fractals", *RAIRO Theoretical Informatics and Applications*, vol. 41, no. 3, pp. 329–349, 2007.

[BER 07d]  BERTHÉ V., RIGO M., "Odometers on regular languages", *Theory of Computing Systems,* vol. 40, no. 1, pp. 1–31, 2007.

[BER 08]  BERNAT J., "Symmetrized $\beta$-integers", *Theoretical Computer Science,* vol. 391, nos. 1–2, pp. 164–177, 2008.

[BER 10]  BERTHÉ V., RIGO M. (eds.), *Combinatorics, automata and number theory*, Encyclopedia of Mathematics and Its Applications, Cambridge University Press, Cambridge, vol. 135, 2010.

[BER 11]  BERSTEL J., REUTENAUER C., *Noncommutative rational series with applications*, Encyclopedia of Mathematics and Its Applications, Cambridge University Press, Cambridge, vol. 137, 2011.

[BER 12]  BERTHÉ V., "Numeration and discrete dynamical systems", *Computing,* vol. 94, nos. 2–4, pp. 369–387, 2012.

[BER 14]  BERTHÉ V., DELECROIX V., *S-adic Expansions: A Combinatorial, Arithmetic and Geometric Approach,* RIMS Lecture Note, (in press), Kokyuroku Bessatu, Kyoto University, 2014.

[BÈS 97]  BÈS A., "Undecidable extensions of Büchi arithmetic and Cobham-Semënov theorem", *Journal of Symbolic Logic,* vol. 62, pp. 1280–1296, 1997.

[BÈS 00]  BÈS A., "An extension of the Cobham-Semënov theorem", *Journal of Symbolic Logic,* vol. 65, no. 1, pp. 201–211, 2000.

[BÈS 01]  BÈS A., "A survey of arithmetical definability", *Bulletin of the Belgian Mathematical Society- Simon Stevin*, supplement, pp. 1–54, 2001.

[BLA 14]  BLANCHET-SADRI F., CURRIE J.D., RAMPERSAD N., FOX N., "Abelian complexity of fixed point of morphism $0 \mapsto 012, 1 \mapsto 02, 2 \mapsto 1$", *Integers,* vol. 14, A11, 2014.

[BLO 02] BLONDEL V.D., PORTIER N., "The presence of a zero in an integer linear recurrent sequence is NP-hard to decide", *Linear Algebra and Its Applications*. vol. 351/352, pp. 91–98, 2002.

[BLO 09] BLONDEL V.D., CASSAIGNE J., JUNGERS R.M., "On the number of $\alpha$-power-free binary words for $2 < \alpha \leq 7/3$", *Theoretical Computer Science*, vol. 410, nos. 30–32, pp. 2823–2833, 2009.

[BOI 98] BOIGELOT B., RASSART S., WOLPER P., "On the expressiveness of real and integer arithmetic automata (extended abstract)", in *ICALP*, Lecture Notes in Computer Science, Springer-Verlag, vol. 1443, pp. 152–163, 1998.

[BOI 04] BOIGELOT B., LATOUR L., "Counting the solutions of Presburger equations without enumerating them", *Theoretical Computer Science*, vol. 313, no. 1, pp. 17–29, 2004.

[BOI 08] BOIGELOT B., BRUSTEN J., BRUYÈRE V., "On the sets of real numbers recognized by finite automata in multiple bases", *Proceeding of 35th ICALP (Reykjavik)*, Lecture Notes in Computer Science, Springer-Verlag, vol. 5126, pp. 112–123, 2008.

[BOI 09a] BOIGELOT B., BRUSTEN J., "A generalization of Cobham's theorem to automata over real numbers", *Theoretical Computer Science*, vol. 410, no. 18, pp. 1694–1703, 2009.

[BOI 09b] BOIGELOT B., BRUSTEN J., LEROUX J., "A generalization of Semenov's theorem to automata over real numbers", in SCHMIDT R.A. (ed.), *Automated Deduction, 22nd International Conference, CADE*, McGill University, Montreal, Lecture Notes in Computer Science, vol. 5663, pp. 469–484, 2009.

[BOR 09] BOREL É., "Les probabilités dénombrables et leurs applications arithmétiques", *Rendiconti del Circolo Matematico di Palermo*, vol. 27, pp. 247–271, 1909.

[BOY 89] BOYD D.W., "Salem numbers of degree four have periodic expansions", in *Théorie des Nombres (Québec, PQ, 1987)*, de Gruyter, Berlin, pp. 57–64, 1989.

[BOY 96] BOYD D.W., "On beta expansions for Pisot numbers", *Mathematics of Computation*, vol. 65, no. 214, pp. 841–860, 1996.

[BRA 83] BRANDENBURG F.-J., "Uniformly growing $k$-th power-free homomorphisms", *Theoretical Computer Science*, vol. 23, pp. 69–82, 1983.

[BRL 89] BRLEK S., "Enumeration of factors in the Thue-Morse word", *Discrete Applied Mathematics*, vol. 24, pp. 83–96, 1989.

[BRL 06] BRLEK S., DULUCQ S., LADOUCEUR A., VUILLON L., "Combinatorial properties of smooth infinite words", *Theoretical Computer Science*, vol. 352, nos. 1–3, pp. 306–317, 2006.

[BRL 07] BRLEK S., MELANÇON G., PAQUIN G., "Properties of the extremal infinite smooth words", *Discrete Mathematics & Theoretical Computer Science*, vol. 9, no. 2, pp. 33–49, 2007.

[BRU 46] DE BRUIJN N.G., "A combinatorial problem", *Proceedings of the Nederlandse Akademie van Wetenschappen*, vol. 49, pp. 758–764, 1946.

[BRU 75] DE BRUIJN N.G., "Acknowledgement of priority to C. Flye Sainte-Marie on the counting of circular arrangements of $2n$ zeros and ones that show each $n$-letter word exactly once", Technical Report no. 75-WSK-06, Department of Mathematics and Computing Science, Eindhoven University of Technology, The Netherlands, June 1975.

[BRU 81] DE BRUIJN N.G., *Asymptotic Methods in Analysis*, Dover, 1981.

[BRU 94] BRUYÈRE V., HANSEL G., MICHAUX C., VILLEMAIRE R., "Logic and $p$-recognizable sets of integers", *Bulletin of the Belgian Mathematical Society*, vol. 1, pp. 191–238, 1994.

[BRU 95] BRUYÈRE V., HANSEL G., "Recognizable sets of numbers in nonstandard bases", in BAEZA-YATES R., GOLES E., POBLETE P.V. (eds.), *LATIN '95: Theoretical Informatics*, Lecture Notes in Computer Science, Springer-Verlag, vol. 911, pp. 167–179, 1995.

[BRU 97] BRUYÈRE V., HANSEL G., "Bertrand numeration systems and recognizability", *Theoretical Computer Science*, vol. 181, pp. 17–43, 1997.

[BRU 11] BRUALDI R.A., "The mutually beneficial relationship of graphs and matrices", *CBMS Regional Conference Series in Mathematics*, American Mathematical Society, Providence, RI, vol. 115, 2011.

[BÜC 60] BÜCHI J.R., "Weak secord-order arithmetic and finite automata", *Zeitschrift für mathematische Logik und Grundlagen der Mathematik*, vol. 6, pp. 66–92, 1960.

[BUC 09a] BUCCI M., DE LUCA A., DE LUCA A., "Characteristic morphisms of generalized episturmian words", *Theoretical Computer Science*, vol. 410, nos. 30–32, pp. 2840–2859, 2009.

[BUC 09b] BUCCI M., DE LUCA A., DE LUCA A., ZAMBONI L.Q., "On θ-episturmian words", *European Journal of Combinatorics*, vol. 30, no. 2, pp. 473–479, 2009.

[BUG 12] BUGEAUD Y., *Distribution modulo one and Diophantine approximation, Cambridge Tracts in Mathematics*, Cambridge University Press, Cambridge, vol. 193, 2012.

[BUR 98] BURDÍK Č., FROUGNY C., GAZEAU J.-P., KREJCAR R., "Beta-integers as natural counting systems for quasicrystals", *Journal of Physics A*, vol. 31, no. 30, pp. 6449–6472, 1998.

[CAN 01] CANTERINI V., SIEGEL A., "Automate des préfixes-suffixes associé à une substitution primitive", *Journal de Théorie des Nombres de Bordeaux*, vol. 13, no. 2, vol. 353–369, 2001.

[CAR 93] CARPI A., "Repetitions in the Kolakovski sequence", *Bulletin of the European Association for Theoretical Computer Science*, vol. 50, pp. 194–196, 1993.

[CAR 94] CARPI A., "On repeated factors in $C^\infty$-words", *Information Processing Letters*, vol. 52, pp. 289–294, 1994.

[CAR 01] CARPI A., MAGGI C., "On synchronized sequences and their separators", *RAIRO Theoretical Informatics and Applications*, vol. 35, no. 6, pp. 513–524, 2001.

[CAR 07] CARPI A., "On Dejean's conjecture over large alphabets", *Theoretical Computer Science*, vol. 385, nos. 1–3, pp. 137–151, 2007.

[CAR 09] CARPI A., D'ALONZO V., "On the repetitivity index of infinite words", *International Journal of Algebra and Computation*, vol. 19, no. 2, pp. 145–158, 2009.

[CAR 10] CARPI A., D'ALONZO V., "On factors of synchronized sequences", *Theoretical Computer Science*, vol. 411, nos. 44–46, pp. 3932–3937, 2010.

[CAS 93] CASSAIGNE J., "Counting overlap-free binary words", in ENJALBERT P., FINKEL A., WAGNER K.W. (eds.), *STACS 93, Proceeding of 10th Symposium on Theoretical Aspects of Computer Science*, Lecture Notes in Computer Science, Springer-Verlag, vol. 665, pp. 216–225, 1993.

[CAS 95] CASSAIGNE J., KARHUMÄKI J., "Toeplitz words, generalized periodicity and periodically iterated morphisms", Technical Report, LITP, Institut Blaise Pascal, no. 36, 1995.

[CAS 96] CASSAIGNE J., "Special factors of sequences with linear subword complexity", in DASSOW J., ROZENBERG G., SALOMAA A. (eds.), *Developments in Language Theory II*, World Scientific, pp. 25–34, 1996.

[CAS 97a] CASSAIGNE J., "Complexité et facteurs spéciaux", *Bulletin of the Belgian Mathematical Society*, vol. 4, pp. 67–88, 1997.

[CAS 97b] CASSAIGNE J., KARHUMÄKI J., "Toeplitz words, generalized periodicity and periodically iterated morphisms", *European Journal of Combinatorics*, vol. 18, pp. 497–510, 1997.

[CAS 00] CASSAIGNE J., FERENCZI S., ZAMBONI L.Q., "Imbalances in Arnoux-Rauzy sequences", *Annales de l'Institut Fourier (Grenoble)*, vol. 50, pp. 1265–1276, 2000.

[CAS 03] CASSAIGNE J., NICOLAS F., "Quelques propriétés des mots substitutifs", *Bulletin of the Belgian Mathematical Society Simon Stevin*, vol. 10, pp. 661–676, 2003.

[CAS 07] CASSAIGNE J., FRID A.E., "On the arithmetical complexity of Sturmian words", *Theoretical Computer Science*, vol. 380, no. 3, pp. 304–316, 2007.

[CAS 11a] CASSAIGNE J., CURRIE J.D., SCHAEFFER L., SHALLIT J., "Avoiding three consecutive blocks of the same size and same sum", 2011. Available at http://arxiv.org/abs/1106.5204

[CAS 11b] CASSAIGNE J., RICHOMME G., SAARI K., ZAMBONI L.Q., "Avoiding Abelian powers in binary words with bounded Abelian complexity", *International Journal of Foundations of Computer Science*, vol. 22, no. 4, pp. 905–920, 2011.

[CHA 33] CHAMPERNOWNE D.G., "The construction of decimals normal in the scale of ten", *Journal of the London Mathematical Society*, pp. 254–260, 1933.

[CHA 08] CHARLIER É., RIGO M., STEINER W., "Abstract numeration systems on bounded languages and multiplication by a constant", *Integers*, vol. 8, A35, 2008.

[CHA 10] CHARLIER É., KÄRKI T., RIGO M., "Multidimensional generalized automatic sequences and shape-symmetric morphic words", *Discrete Mathematics*, vol. 310, nos. 6–7, pp. 1238–1252, 2010.

[CHA 11] CHARLIER É., RAMPERSAD N., RIGO M., WAXWEILER L., "The minimal automaton recognizing $m\mathbb{N}$ in a linear numeration system", *Integers*, vol. 11B, A4, 2011.

[CHA 12] CHARLIER É., RAMPERSAD N., SHALLIT J., "Enumeration and decidable properties of automatic sequences", *International Journal of Foundations of Computer Science*, vol. 23, no. 5, pp. 1035–1066, 2012.

[CHA 13] CHARLIER É., LEROY J., RIGO M., "An analogue of Cobham's theorem for graph directed iterated function systems", 2013. Available at http://arxiv.org/abs/1310.0309

[CHA 14] CHARLIER É., LEROY J., RIGO M., "Asymptotic properties of free monoid morphisms", 2014. [Preprint]

[CHE 01] CHEKHOVA N., HUBERT P., MESSAOUDI A., "Propriétés combinatoires, ergodiques et arithmétiques de la substitution de Tribonacci", *Journal de Théorie des Nombres de Bordeaux*, vol. 13, pp. 371–394, 2001.

[CHE 09] CHEVALLIER N., "Coding of a translation of the two-dimensional torus", *Monatshefte für Mathematik*, vol. 157, no. 2, pp. 101–130, 2009.

[CHO 95] CHOFFRUT C., GOLDWURM W., "Rational transductions and complexity of counting problems", *Mathematical Systems Theory*, vol. 28, no. 5, pp. 437–450, 1995.

[CHR 80] CHRISTOL G., KAMAE T., MENDÈS FRANCE M., RAUZY G., "Suites algébriques, automates et substitutions", *Bulletin de la Société Mathématique de France*, vol. 108, pp. 401–419, 1980.

[COB 68] COBHAM A., "On the Hartmanis-Stearns problem for a class of tag machines", in *IEEE Conference Record of Ninth Annual Symposium on Switching and Automata Theory*, Also appeared as IBM Research Technical Report no. RC-2178, pp. 51–60, August 23 1968.

[COB 69]  COBHAM A., "On the base-dependence of sets of numbers recognizable by finite automata", *Mathematical Systems Theory,* vol. 3, pp. 186–192, 1969.

[COB 72]  COBHAM A., "Uniform tag sequences", *Mathematical Systems Theory,* vol. 6, pp. 164–192, 1972.

[COH 78]  COHEN D.I.A., *Basic Techniques of Combinatorial Theory,* John Wiley & Sons, New York-Chichester-Brisbane, 1978.

[COV 73]  COVEN E.M., HEDLUND G.A., "Sequences with minimal block growth", *Mathematical Systems Theory,* vol. 7, pp. 138–153, 1973.

[CRI 93]  CRISP D., MORAN W., POLLINGTON A., SHIUE P., "Substitution invariant cutting sequences", *Journal de Théorie des Nombres de Bordeaux,* vol. 5, pp. 123–137, 1993.

[CRO 07]  CROCHEMORE M., HANCART C., LECROQ T., *Algorithms on Strings,* Cambridge University Press, Cambridge, 2007.

[CUL 92]  CULIK II K., KARHUMÄKI J., LEPISTÖ A., "Alternating iteration of morphisms and the Kolakovski [sic] sequence", in ROZENBERG G., SALOMAA A. (eds.), *Lindenmayer Systems,* Springer-Verlag, pp. 93–103, 1992.

[CUR 02]  CURRIE J.D., "There are ternary circular square-free words of length $n$ for $n \geq 18$", *Electronic Journal of Combinatorics,* vol. 9, no. 1, p. 10, 2002.

[CUR 09]  CURRIE J.D., ABERKANE A., "A cyclic binary morphism avoiding abelian fourth powers", *Theoretical Computer Science,* vol. 410, no. 1, pp. 44–52, 2009.

[CUR 11]  CURRIE J., RAMPERSAD N., "A proof of Dejean's conjecture", *Mathematics of Computation,* vol. 80, no. 274, pp. 1063–1070, 2011.

[CUR 12]  CURRIE J.D., RAMPERSAD N., "Fixed points avoiding Abelian $k$-powers", *Journal of Combinatorial Theory, Series A,* vol. 119, no. 5, pp. 942–948, 2012.

[CYR 12]  CYR V., KRA B., "Nonexpansive $Z^2$ subdynamics and Nivat's conjecture", 2012. Available at http://arxiv.org/abs/1208.4090

[CZE 07] CZERWIŃSKI S., GRYTCZUK J., "Nonrepetitive colorings of graphs", in *6th Czech-Slovak International Symposium on Combinatorics, Graph Theory, Algorithms and Applications, Electronic Notes in Discrete Mathematics*, Elsevier, Amsterdam, vol. 28, pp. 453–459, 2007.

[DAM 06] DAMANIK D., LENZ D., "Substitution dynamical systems: characterization of linear repetitivity and applications", *Journal of Mathematical Analysis and Applications*, vol. 321, no. 2, pp. 766–780, 2006.

[DAR 00] DARTNELL P., DURAND F., MAASS A., "Orbit equivalence and Kakutani equivalence with Sturmian subshifts", *Studia Mathematica*, vol. 142, no. 1, pp. 25–45, 2000.

[DAU 06] DAUBECHIES I., YILMAZ Ö., "Robust and practical analog-to-digital conversion with exponential precision", *IEEE Transaction on Informations Theory*, vol. 52, no. 8, pp. 3533–3545, 2006.

[DAU 10] DAUBECHIES I., GÜNTÜRK S., WANG Y., YILMAZ Ö., "The golden ratio encoder", *IEEE Transactions on Information Theory*, vol. 56, no. 10, pp. 5097–5110, 2010.

[DEJ 72] DEJEAN F., "Sur un théorème de Thue", *Journal of Combinatorial Theory, Series A*, vol. 13, pp. 90–99, 1972.

[DEK 79] DEKKING F.M., "Strongly non-repetitive sequences and progression-free sets", *Journal of Combinatorial Theory, Series A*, vol. 27, pp. 181–185, 1979.

[DEK 97] DEKKING F.M., "What is the long range order in the Kolakoski sequence?", in MOODY R.V. (ed.), *The Mathematics of Long-Range Aperiodic Order, NATO ASI Series, Series C., Mathematical and Physical Science*, Kluwer, vol. 489, pp. 115–125, 1997.

[DEN 76] DENKER M., GRILLENBERGER C., SIGMUND K., *Ergodic Theory on Compact Spaces*, Lecture Notes in Mathematics, Springer-Verlag, Berlin, vol. 527, 1976.

[DER 07] DERKSEN H., "A Skolem-Mahler-Lech theorem in positive characteristic and finite automata", *Inventiones Mathematicae*, vol. 168, no. 1, pp. 175–224, 2007.

[DEV 08] DEVYATOV R., "On subword complexity of morphic sequences", *Proceeding of 3rd International Computer Science Symposium in Russia (CSR 2008)*, Lecture Notes in Computer Science, Springer-Verlag, vol. 5010, pp. 146–157, 2008.

[DID 98a] DIDIER G., "Combinatoire des codages des rotations", *Acta Arithmetica,* vol. 85, pp. 157–177, 1998.

[DID 98b] DIDIER G., "Codage de rotations et fractions continues", *Journal of Number Theory,* vol. 71, pp. 275–306, 1998.

[DOM 11] DOMBEK D., MASÁKOVÁ Z., PELANTOVÁ E., "Number representation using generalized $(-\beta)$-transformation", *Theoretical Computer Science,* vol. 412, no. 48, pp. 6653–6665, 2011.

[DRM 97] DRMOTA M., TICHY R.F., *Sequences, Discrepancies, and Applications*, Lecture Notes in Mathematics, Springer-Verlag, vol. 1651, 1997.

[DRO 01] DROUBAY X., JUSTIN J., PIRILLO G., "Episturmian words and some constructions of de Luca and Rauzy", *Theoretical Computer Science,* vol. 255, pp. 539–553, 2001.

[DRO 09] DROSTE M., KUICH W., VOGLER H., (eds.), *Handbook of Weighted Automata*, Monographs in Theoretical Computer Science. An EATCS Series, Springer-Verlag, Berlin, 2009.

[DUC 08a] DUCHÊNE E., RIGO M., "Cubic Pisot unit combinatorial games", *Monatshefte für Mathematik,* vol. 155, nos. 3–4, pp. 217–249, 2008.

[DUC 08b] DUCHÊNE E., RIGO M., "A morphic approach to combinatorial games: the Tribonacci case", *RAIRO Theoretical Informatics and Applications* , vol. 42, no. 2, pp. 375–393, 2008.

[DUC 10a] DUCHÊNE E., FRAENKEL A.S., NOWAKOWSKI R.J., RIGO M., "Extensions and restrictions of Wythoff's game preserving its $P$ positions", *Journal of Combinatorial Theory, Series A,* vol. 117, no. 5, pp. 545–567, 2010.

[DUC 10b] DUCHÊNE E., RIGO M., "Invariant games", *Theoretical Computer Science,* vol. 411, nos. 34–36, pp. 3169–3180, 2010.

[DUM 89] DUMONT J.-M., THOMAS A., "Systèmes de numération et fonctions fractales relatifs aux substitutions", *Theoretical Computer Science,* vol. 65, pp. 153–169, 1989.

[DUR 98a] DURAND F., "A generalization of Cobham's theorem", *Theory of Computing Systems*, vol. 31, pp. 169–185, 1998.

[DUR 98b] DURAND F., "A characterization of substitutive sequences using return words", *Discrete Mathematics*, vol. 179, pp. 89–101, 1998.

[DUR 98c] DURAND F., "Sur les ensembles d'entiers reconnaissables", *Journal de Théorie des Nombres de Bordeaux*, vol. 10, pp. 65–84, 1998.

[DUR 99] DURAND F., HOST B., SKAU C., "Substitutional dynamical systems, Bratteli diagrams and dimension groups", *Ergodic Theory and Dynamical Systems*, vol. 19, no. 4, pp. 953–993, 1999.

[DUR 00] DURAND F., "Linearly recurrent subshifts have a finite number of non-periodic subshift factors", *Ergodic Theory and Dynamical Systems*, vol. 20, no. 4, pp. 1061–1078, 2000.

[DUR 02] DURAND F., "A theorem of Cobham for non-primitive substitutions", *Acta Arithmetica*, vol. 104, no. 3, pp. 225–241, 2002.

[DUR 03] DURAND F., "Corrigendum and addendum to: 'Linearly recurrent subshifts have a finite number of non-periodic subshift factors'", *Ergodic Theory Dynamical Systems*, vol. 23, no. 2, pp. 63–669, 2003.

[DUR 09] DURAND F., RIGO M., "Syndeticity and independent substitutions", *Advances in Applied Mathematics*, vol. 42, no. 1, pp. 1–22, 2009.

[DUR 11] DURAND F., "Cobham's theorem for substitutions", *Journal of the European Mathematical Society (JEMS)*, vol. 13, no. 6, pp. 1799–1814, 2011.

[DUR 12] DURAND F., "HD0L $\omega$-equivalence and periodicity problems in the primitive case", *Uniform Distribution Theory*, vol. 7, no. 1, pp. 199–215, 2012.

[DUR 13a] DURAND F., "Decidability of the HD0L ultimate periodicity problem", *RAIRO Theoretical Informatics and Applications*, vol. 47, no. 2, pp. 201–214, 2013.

[DUR 13b] DURAND F., RIGO M., "Multidimensional extension of the Morse-Hedlund theorem", *European Journal of Combinatorics*, vol. 34, no. 2, pp. 391–409, 2013.

[EBB 94] EBBINGHAUS H.-D., FLUM J., THOMAS W., *Mathematical Logic*, Undergraduate Texts in Mathematics, 2nd Ed., Springer-Verlag, New York, NY, 1994.

[EIL 74] EILENBERG S., *Automata, Languages, and Machines*, vol. A, Academic Press, 1974.

[END 01] ENDERTON H.B., *A Mathematical Introduction to Logic*, 2nd Ed., Harcourt/Academic Press, Burlington, MA, 2001.

[ENG 98] ENGEL A., *Problem-Solving Strategies*, Problem Books in Mathematics, Springer-Verlag, New York, 1998.

[ENT 74] ENTRINGER R.C., JACKSON D.E., SCHATZ J.A., "On nonrepetitive sequences", *Journal of Combinatorial Theory, Series A*, vol. 16, pp. 159–164, 1974.

[ERD 61] ERDÖS P., "Some unsolved problems", *Magyar Tud. Akad. Mat. Kutató Int. Közl.*, vol. 6, pp. 221–254, 1961.

[FAG 97] FAGNOT I., "Sur les facteurs des mots automatiques", *Theoretical Computer Science*, vol. 172, pp. 67–89, 1997.

[FEK 23] FEKETE M., "Über die verteilung der wurzeln bei gewissen algebraischen gleichungen mit ganzzahligen koeffizienten", *Mathematische Zeitschrift*, vol. 17, pp. 228–249, 1923.

[FER 96] FERENCZI S., "Rank and symbolic complexity", *Ergodic Theory and Dynamical Systems*, vol. 16, pp. 663–682, 1996.

[FER 03] FERENCZI S., HOLTON C., ZAMBONI L.Q., "Structure of three-interval exchange transformations. II. A combinatorial description of the trajectories", *Journal d'Analyse Mathématique*, vol. 89, pp. 239–276, 2003.

[FER 06] FERENCZI S., "Substitution dynamical systems on infinite alphabets", *Annales de l'Institut Fourier (Grenoble)*, vol. 56, no. 7, pp. 2315–2343, 2006.

[FER 07] FERNIQUE T., "Local rule substitutions and stepped surfaces", *Theoretical Computer Science*, vol. 380, no. 3, pp. 317–329, 2007.

[FER 08] FERENCZI S., ZAMBONI L.Q., "Languages of $k$-interval exchange transformations", *Bulletin of the London Mathematical Society*, vol. 40, no. 4, pp. 705–714, 2008.

[FIC 13] FICI G., LANGIU A., LECROQ T., LEFEBVRE A., MIGNOSI F., PRIEUR-GASTON E., "Abelian repetitions in sturmian words", *DLT, Marne-la-Vallée*, Lecture Notes in Computer Science, Springer-Verlag, vol. 7907, pp. 227–238, 2013.

[FIN 65] FINE N.J., WILF H.S., "Uniqueness theorems for periodic functions", *Proceedings of the American Mathematical Society*, vol. 16, pp. 109–114, 1965.

[FIS 74] FISCHER M.J., RABIN M.O., "Super-exponential complexity of Presburger arithmetic", *Complexity of Computation (Proceedings SIAM-AMS Symposium, New York, 1973)*, pp. 27–41, American Mathematical Society, Providence, RI, vol. VII, 1974.

[FIS 06] FISCHLER S., "Palindromic prefixes and episturmian words", *Journal of Combinatorial Theory, Series A*, vol. 113, no. 7, pp. 1281–1304, 2006.

[FLO 09] FLOREK J., "Billiards and the five distance theorem", *Acta Arithmetica*, vo. 139, no. 3, pp. 229–239, 2009.

[FLY 94] FLYE SAINTE-MARIE C., "Question 48", *L'Intermédiaire des Mathématiciens*, vol. 1, pp. 107–110, 1894.

[FON 07] FON-DER-FLAASS D.G., FRID A.E., "On periodicity and low complexity of infinite permutations", *European Journal of Combinatorics*, vol. 28, no. 8, pp. 2106–2114, 2007.

[FRA 73] FRAENKEL A.S., "Complementing and exactly covering sequences", *Journal of Combinatorial Theory, Series A*, vol. 14, pp. 8–20, 1973.

[FRA 82] FRAENKEL A.S., "How to beat your Wythoff games' opponent on three fronts", *American Mathematical Monthly*, vol. 89, no. 6, pp. 353–361, 1982.

[FRA 85] FRAENKEL A.S., "Systems of numeration", *American Mathematical Monthly*, vol. 92, pp. 105–114, 1985.

[FRA 98] FRAENKEL A.S., SIMPSON J., "How many squares can a string contain?", *Journal of Combinatorial Theory, Series A*, vol. 82, pp. 112–120, 1998.

[FRA 10] FRAENKEL A.S., "Complementary iterated floor words and the Flora game", *SIAM Journal on Discrete Mathematics*, vol. 24, no. 2, pp. 570–588, 2010.

[FRI 98a] FRID A., "On uniform D0L words", MORVAN M., MEINEL C., KROB D., (eds.), *STACS '98, Proceedings of the 15th Symposium on Theoretical Aspects of Computer Science*, Lecture Notes in Computer Science, Springer-Verlag, vol. 1373, pp. 544–554, 1998.

[FRI 98b] FRID A.E., "On the frequency of factors in a D0L word", *Journal of Automata, Languages, and Combinatorics*, vol. 3, pp. 29–41, 1998.

[FRI 99] FRID A.E., AVGUSTINOVICH S.V., "On bispecial words and subword complexity of D0L sequences", DING C., HELLESETH T., NIEDERREITER H., (eds.), *Sequences and Their Applications, Proceedings of SETA '98*, Springer-Verlag, pp. 191–204, 1999.

[FRI 05] FRID A.E., "Sequences of linear arithmetical complexity", *Theoretical Computer Science*, vol. 339, no. 1, pp. 68–87, 2005.

[FRI 06] FRID A.E., "On possible growths of arithmetical complexity", *RAIRO Theoretical Informatics and Applications*, vol. 40, no. 3, pp. 443–458, 2006.

[FRI 12a] FRID A.E., "Fine and Wilf's theorem for permutations", *Sibirskie lektronnye Matematicheskie Izvestiya*, vol. 9, pp. 377–381, 2012.

[FRI 12b] FRID A., ZAMBONI L., "On automatic infinite permutations", *RAIRO Theoretical Informatics and Applications*, vol. 46, no. 1pp. 77–85, 2012.

[FRI 13] FRID A., JAMET D., "The number of binary rotation words" 2013. Available at http://arxiv.org/abs/1302.3722

[FRO 92a] FROUGNY C., "Representations of numbers and finite automata", *Mathematical Systems Theory*, vol. 25, pp. 37–60, 1992a1992.

[FRO 92b] FROUGNY C., SOLOMYAK B., "Finite beta-expansions, *Ergodic Theory and Dynamical Systems*, vol. 12, pp. 713–723, 1992.

[FRO 93] FROUGNY C., SAKAROVITCH J., "Synchronized rational relations of finite and infinite words", *Theoretical Computer Science*, vol. 108, no. 1, pp. 45–82, 1993.

[FRO 96a] FROUGNY C., "On the successor function in non-classical numeration systems", PUECH C., REISCHUK R., (eds.), *STACS '96, Proceedings of the 13th Symposium on Theoretical Aspects of Computer Science*, Lecture Notes in Computer Science, Springer-Verlag, vol. 1046, pp. 543–553, 1996.

[FRO 96b] FROUGNY C., SOLOMYAK B., "On representation of integers in linear numeration systems", POLLICOTT M., SCHMIDT K., (eds.), *Ergodic Theory of $\mathbb{Z}^d$ Actions (Warwick, 1993–1994)*, London Mathematical Society Lecture Note Series, Cambridge University Press, vol. 228, pp. 345–368, 1996.

[FRO 97] FROUGNY C., "On the sequentiality of the successor function", *Information and Computation*, vol. 139, no. 1, pp. 17–38, 1997.

[FRO 03] FROUGNY C., GAZEAU J.-P., KREJCAR R., "Additive and multiplicative properties of point sets based on beta-integers", *Theoretical Computer Science*, vol. 303, nos. 2–3, pp. 491–516, 2003.

[FRO 07] FROUGNY C., "Non-standard number representation: computer arithmetic, beta-numeration and quasicrystals", *Physics and Theoretical Computer Science, NATO Science for Peace and Security Series D: Information and Communication Security*, IOS, Amsterdam, vol. 7, pp. 155–169, 2007.

[FRO 11] FROUGNY C., PELANTOVÁ E., SVOBODOVÁ M., "Parallel addition in non-standard numeration systems", *Theoretical Computer Science*, vol. 412, no. 41, pp. 5714–5727, 2011.

[FUR 67] FURSTENBERG H., "Algebraic functions over finite fields", *Journal of Algebra*, vol. 7, pp. 271–277, 1967.

[FUR 81] FURSTENBERG H., *Recurrence in Ergodic Theory and Combinatorial Number Theory*, Princeton University Press, 1981.

[GAN 60] GANTMACHER F.R., *The Theory of Matrices*, Chelsea, 1960.

[GAW 10] GAWRYCHOWSKI P., KRIEGER D., RAMPERSAD N., SHALLIT J., "Finding the growth rate of a regular or context-free language in polynomial time", *International Journal of Foundations of Computer Science*, vol. 21, no. 4, pp. 597–618, 2010.

[GEL 59]  GELFOND A.O., "A common property of number systems", *Izvestiya Akademii Nauk SSSR. Seriya Matematicheskaya*, vol. 23, pp. 809–814, 1959.

[GIL 81]  GILBERT W.J., "Radix representations of quadratic fields", *Journal of Mathematical Analysis and Applications*, vol. 83, pp. 264–274, 1981.

[GIN 64]  GINSBURG S., SPANIER E.H., "Bounded ALGOL-like languages", *Transactions of the American Mathematical Society*, vol. 113, pp. 333–368, 1964.

[GLE 09]  GLEN A., JUSTIN J., "Episturmian words: a survey", *RAIRO Theoretical Informatics and Applications*, vol. 43, no. 3, pp. 403–442, 2009.

[GOČ 12]  GOČ D., HENSHALL D., SHALLIT J., "Automatic theorem-proving in combinatorics on words", *Implementation and Application of Automata*, Lecture Notes in Computer Science, Springer-Verlag, vol. 7381, pp. 180–191, 2012.

[GOČ 13a]  GOČ D., RAMPERSAD N., RIGO M., SALIMOV P., "On the number of abelian bordered words (with an example of automatic theorem-proving)", 2013. [Preprint]

[GOČ 13b]  GOČ D., MOUSAVI H., SHALLIT J., "On the number of unbordered factors", *Language and Automata Theory and Applications*, Lecture Notes in Computer Science, Springer-Verlag, vol. 7810, pp. 299–310, 2013.

[GOČ 13c]  GOČ D., SCHAEFFER L., SHALLIT J., "Subword complexity and $k$-synchronization", *DLT 2013, Marne-la-Vallée*, Lecture Notes in Computer Science, Springer-Verlag, vol. 7907, pp. 252–263, 2013.

[GOD 01]  GODSIL C., ROYLE G., *Algebraic Graph Theory*, Graduate Texts in Mathematics, Springer-Verlag, New York, vol. 207, 2001.

[GOL 65]  GOLOMB S.W., BAUMERT L.D., "Backtrack programming", *Journal of the Association for Computing Machinery*, vol. 12, pp. 516–524, 1965.

[GRA 89]  GRAHAM R.L., KNUTH D.E., PATASHNIK O., *Concrete Mathematics*, Addison-Wesley, 1989.

[GRA 95]  GRABNER P.J., LIARDET P., TICHY R.F., "Odometers and systems of numeration", *Acta Arithmetica*, vol. 70, pp. 103–123, 1995.

[GRA 06] GRABNER P., "Uniform distribution, digital functions, and applications", Lecture Notes CANT Summer School, University of Liège, May 2006.

[GRE 14] GREINECKER F., "On the 2-abelian complexity of thue-morse subwords", 2014. Available at http://arxiv.org/abs/1405.3906

[GRY 07a] GRYTCZUK J., "Nonrepetitive graph coloring", *Graph Theory in Paris*, Trends in Mathematics, Birkhäuser, Basel, pp. 209–218, 2007.

[GRY 07b] GRYTCZUK J., "Pattern avoidance on graphs", *Discrete Mathematics*, vol. 307, nos. 11–12, pp. 1341–1346, 2007.

[GUA 09] GUAY-PAQUET M., SHALLIT J., "Avoiding squares and overlaps over the natural numbers", *Discrete Mathematics*, vol. 309, no. 21, pp. 6245–6254, 2009.

[GÜN 12] GÜNTÜRK C.S., "Mathematics of analog-to-digital conversion", *Communications on Pure and Applied Mathematics*, vol. 65, no. 12, pp. 1671–1696, 2012.

[HAL 01] HALAVA V., HARJU T., "Mortality in matrix semigroups", *American Mathematical Monthly*, vol. 108, no. 7, pp. 649–653, 2001.

[HAN 86] HANSEL G., "Une démonstration simple du théorème de Skolem-Mahler-Lech", *Theoretical Computer Science*, vol. 43, no.1, pp. 91–98, 1986.

[HAR 66] HARTMANIS J., STEARNS R.E., *Algebraic Structure Theory of Sequential Machines*, Prentice-Hall, Englewood Cliffs, NJ, 1966.

[HAR 85] HARDY G.H., WRIGHT E.M., *An Introduction to the Theory of Numbers*, 5th ed., Oxford University Press, 1985.

[HAR 86] HARJU T., LINNA M., "On the periodicity of morphisms on free monoids", *RAIRO Informatique Théorique et Applications*, vol. 20, pp. 47–54, 1986.

[HED 69] HEDLUND G.A., "Endomorphisms and automorphisms of the shift dynamical system", *Mathematical Systems Theory*, vol. 3, pp. 320–375, 1969.

[HOD 83] HODGSON B., "Décidabilité par automate fini", *Annales des Sciences Mathématiques du Québec*, vol. 7, pp. 39–57, 1983.

[HOL 98] HOLLANDER M., "Greedy numeration systems and regularity", *Theory of Computer Systems,* vol. 31, pp. 111–133, 1998.

[HOL 99] HOLTON C., ZAMBONI L.Q., "Descendants of primitive substitutions", *Theory of Computer Systems,* vol. 32, pp. 133–157, 1999.

[HOL 01] HOLTON C., ZAMBONI L.Q., "Directed graphs and substitutions", *Theory of Computer Systems,* vol. 34, pp. 545–564, 2001.

[HOL 13] HOLUB Š., "Abelian powers in paper-folding words", *Journal of Combinatorial Theory, Series A,* vol. 120, no. 4, pp. 872–881, 2013.

[HON 84] HONKALA J., "Bases and ambiguity of number systems", *Theoretical Computer Science,* vol. 31, pp. 61–71, 1984.

[HON 86] HONKALA J., "A decision method for the recognizability of sets defined by number systems", *RAIRO Informatique Théorique et Applications,* vol. 20, pp. 395–403, 1986.

[HON 92] HONKALA J., "On unambiguous number systems with a prime power base", *Acta Cybernetica,* vol. 10, pp. 155–163, 1992.

[HON 09] HONKALA J., "On the simplification of infinite morphic words", *Theoretical Computer Science,* vol. 410, nos. 8–10, pp. 997–1000, 2009.

[HOP 79] HOPCROFT J.E., ULLMAN J.D., *Introduction to Automata Theory, Languages, and Computation,* Addison-Wesley, 1979.

[HUO 12] HUOVA M., KARHUMÄKI J., SAARELA A., "Problems in between words and abelian words: $k$-abelian avoidability", *Theoretical Computer Science,* vol. 454, pp. 172–177, 2012.

[HUO 13] HUOVA M., KARHUMÄKI J., "On the unavoidability of $k$-abelian squares in pure morphic words", *Journal of Integer Sequences,* vol. 16, no. 2, Article 13.2.9, 2013.

[ILI 05] ILIE L., OCHEM P., SHALLIT J., "A generalization of repetition threshold", *Theoretical Computer Science,* vol. 345, nos. 2–3, pp. 359–369, 2005.

[ITO 69] ITO R., "Every semilinear set is a finite union of disjoint linear sets", *Journal of Computer and System Sciences,* vol. 3, pp. 221–231, 1969.

[ITO 09] ITO S., SADAHIRO T., "Beta-expansions with negative bases", *Integers*, vol. 9, no. A22, pp. 239–259, 2009.

[JUN 08] JUNGERS R.M., PROTASOV V., BLONDEL V.D., "Efficient algorithms for deciding the type of growth of products of integer matrices", *Linear Algebra and its Applications*, vol. 428, no. 10, pp. 2296–2311, 2008.

[JUS 00] JUSTIN J., VUILLON L., "Return words in Sturmian and episturmian words", *Theoretical Informatics and Applications*, vol. 34, pp. 343–356, 2000.

[JUS 02] JUSTIN J., PIRILLO G., "Episturmian words and episturmian morphisms", *Theoretical Computer Science*, vol. 276, nos. 1–2, pp. 281–313, 2002.

[KAM 02] KAMAE T., ZAMBONI L., "Sequence entropy and the maximal pattern complexity of infinite words", *Ergodic Theory and Dynamical Systems*, vol. 22, no. 4, pp. 1191–1199, 2002.

[KAN 10] KANEL-BELOV A.Y., CHERNYAT'EV A.L., "Describing the set of words generated by interval exchange transformation", *Communications in Algebra*, vol. 38, no. 7, pp. 2588–2605, 2010.

[KAO 08] KAO J.-Y., SHALLIT J., XU Z., "The Frobenius problem in a free monoid", *STACS 2008: 25th International Symposium on Theoretical Aspects of Computer Science, Leibniz International Proceedings in Informatics (LIPIcs)*, Schloss Dagstuhl. Leibniz-Zentrum für Informatik, Wadern, vol. 1, pp. 421–432, 2008.

[KAR 80] KARHUMÄKI J., "Generalized Parikh mappings and homomorphisms", *Information and Control*, vol. 47, no. 3, pp. 155–165, 1980.

[KAR 04] KARHUMÄKI J., SHALLIT J., "Polynomial versus exponential growth in repetition-free binary words", *Journal of Combinatorial Theory, Series A*, vol. 105, no. 2, pp. 335–347, 2004.

[KAR 13a] KARHUMÄKI J., SAARELA A., ZAMBONI L.Q., "On a generalization of Abelian equivalence and complexity of infinite words", *Journal of Combinatorial Theory, Series A*, vol. 120, no. 8, pp. 2189–2206, 2013.

[KAR 13b] KARHUMÄKI J., SAARELA A., ZAMBONI L.Q., "Variations of the Morse-Hedlund theorem for k-abelian equivalence", 2013. Available at http://arxiv.org/abs/1302.3783

[KÁT 81] KÁTAI I., KOVÁCS B., "Canonical number systems in imaginary quadratic fields", *Acta Mathematica Academiae Scientiarum Hungaricae*, vol. 37, pp. 159–164, 1981.

[KEA 75] KEANE M.S., "Interval exchange transformations", *Mathematische Zeitschrift*, vol. 141, pp. 25–31, 1975.

[KER 92] KERÄNEN V., "Abelian squares are avoidable on 4 letters", KUICH W., (ed.), *Proceedings of the 19th International Conference on Automata, Languages, and Programming (ICALP)*, Lecture Notes in Computer Science, Springer-Verlag, vol. 623, pp. 41–52, 1992.

[KER 09] KERÄNEN V., "A powerful abelian square-free substitution over 4 letters", *Theoretical Computer Science*, vol. 410, nos. 38–40, pp. 3893–3900, 2009.

[KES 67] KESTEN H., "On a conjecture of Erdös and Szüsz related to uniform distribution mod 1", *Acta Arithmetica*, vol. 12, pp. 193–212, 1966/1967.

[KHO 01] KHOUSSAINOV B., NERODE A., *Automata Theory and its Applications*, Progress in Computer Science and Applied Logic, Birkhäuser Boston Inc., Boston, MA, vol. 21, 2001.

[KIM 98] KIMBERLING C., "Edouard Zeckendorf", *Fibonacci Quarterly*, vol. 36, pp. 416–418, 1998.

[KIR 13] KIRSCHENHOFER P., THUSWALDNER J., "Shift radix systems - a survey", 2013. Available at http://arxiv.org/abs/1312.0386

[KLA 08] KLAEDTKE F., "Bounds on the automata size for Presburger arithmetic", *ACM Transactions on Computational Logic*, vol. 9, no. 2, Article 11, 2008.

[KLO 12] KLOUDA K., "Bispecial factors in circular non-pushy D0L languages", *Theoretical Computer Science*, vol. 445, pp. 63–74, 2012.

[KNU 60] KNUTH D.E., "An imaginary number system", *Communications of the ACM*, vol. 3, pp. 245–247, 1960.

[KNU 12] KNUTH D.E., *The Art of Computer Programming: Seminumerical Algorithms*, 3rd ed., Addison-Wesley, vol. 2, 2012.

[KOB 98] KOBLITZ N., *Algebraic Aspects of Cryptography*, Algorithms and Computation in Mathematics, Springer-Verlag, Berlin, vol. 3, 1998.

[KOL 65] KOLAKOSKI W., "Elementary problem 5304", *American Mathematical Monthly*, vol. 72, p. 674, 1965.

[KOS 98] KOSKAS M., "Complexité de suites de Toeplitz", *Discrete Mathematics*, vol. 183, pp. 161–183, 1998.

[KRI 09] KRIEGER D., MILLER A., RAMPERSAD N., RAVIKUMAR B., SHALLIT J., "Decimations of languages and state complexity", *Theoretical Computer Science*, vol. 410, nos. 24–25, pp. 2401–2409, 2009.

[KRI 10] KRIEGER D., "The critical exponent of the Arshon words", *RAIRO Theoretical Informatics and Applications*, vol. 44, no. 1, pp. 139–150, 2010.

[LAC 12] LACROIX A., RAMPERSAD N., RIGO M., VANDOMME É., "Syntactic complexity of ultimately periodic sets of integers and application to a decision procedure", *Fundamenta Informaticae*, vol. 116, no. 1–4, pp. 175–187, 2012.

[LAR 11] LARSSON U., HEGARTY P., FRAENKEL A.S., "Invariant and dual subtraction games resolving the Duchêne-Rigo conjecture", *Theoretical Computer Science*, vol. 412, nos. 8–10, pp. 729–735, 2011.

[LAT 06] LATOUR L., Presburger arithmetic: from automata to formulas, PhD Thesis, University of Liege, 2006.

[LAW 04] LAWSON M.V., *Finite Automata*, Chapman & Hall, CRC, Boca Raton, FL, 2004.

[LEC 01] LECOMTE P.B.A., RIGO M., "Numeration systems on a regular language, *Theory of Computing Systems*, vol. 34, pp. 27–44, 2001.

[LEC 02] LECOMTE P., RIGO M., "On the representation of real numbers using regular languages", *Theory of Computing Systems*, vol. 35, no. 1, pp. 13–38, 2002.

[LEC 04] LECOMTE P., RIGO M., "Real numbers having ultimately periodic representations in abstract numeration systems", *Information and Computation*, vol. 192, no. 1, pp. 57–83, 2004.

[LEE 57] LEECH J., "A problem on strings of beads", *Mathematical Gazette*, vol. 41, pp. 277–278, 1957.

[LEG 06] LE GONIDEC M., "Sur la complexité de mots infinis engendrés par des $q$-automates dénombrables", *Annales de l'Institut Fourier (Grenoble)*, vol. 56, no. 7, pp. 2463–2491, 2006.

[LEH 64] LEHMER D.H., "The machine tools of combinatorics", BECKENBACH E.F., (ed.), *Applied Combinatorial Mathematics*, Wiley, New York, pp. 5–31, 1964.

[LEH 93] LEHR S., "Sums and rational multiples of $q$–automatic sequences are $q$–automatic", *Theoretical Computer Science*, vol. 108, pp. 385–391, 1993.

[LEH 96] LEHR S., SHALLIT J., TROMP J., "On the vector space of the automatic reals", *Theoretical Computer Science*, vol. 163, pp. 193–210, 1996.

[LEN 05] LENSTRA H., "Profinite Fibonacci numbers", *Nieuw Archief voor Wiskdunde (5)*, vol. 6, no. 4, pp. 297–300, 2005.

[LEP 93] LEPISTÖ A., "On the power of periodic iteration of morphisms", LINGAS A., KARLSSON R., CARLSSON A., (eds.), *Proceedings of the 20th International Conference on Automata, Languages, and Programming (ICALP)*, Lecture Notes in Computer Science, Springer-Verlag, vol. 700, pp. 496–506, 1993.

[LER 05] LEROUX J., "A polynomial time Presburger criterion and synthesis for number decision diagrams", *Logic in Computer Science*, IEEE Computer Society Press, pp. 147–156, 2005.

[LER 12a] LEROY J., Contribution à la résolution de la conjecture S-adique, PhD Thesis, University de Picardie Jules Verne, January 2012.

[LER 12b] LEROY J., "Some improvements of the $S$-adic conjecture", *Advances in Applied Mathematics*, vol. 48, no. 1, pp. 79–98, 2012.

[LEV 13] LEVITT G., "Subword complexity in free groups", *WORDS 2013 (Turku, Finland)*, Lecture Notes in Computer Science, Springer-Verlag, vol. 8079, p. 14, 2013.

[LI 08] LI M., VITÁNYI P., *An introduction to Kolmogorov Complexity and its Applications*, 3rd ed., Texts in Computer Science, Springer-Verlag, New York, 2008.

[LIA 12] LIAO L., STEINER W., "Dynamical properties of the negative beta-transformation", *Ergodic Theory and Dynamical Systems*, vol. 32, no. 5, pp. 1673–1690, 2012.

[LID 97] LIDL R., NIEDERREITER H., *Finite Fields*, Encyclopedia of Mathematics and Its Applications, 2nd ed., Cambridge University Press, Cambridge, vol. 20, 1997.

[LIN 92] VAN LINT J.L., WILSON R., *A Course in Combinatorics*, Cambridge University Press, 1992.

[LIN 95] LIND D., MARCUS B., *An Introduction to Symbolic Dynamics and Coding*, Cambridge University Press, 1995.

[LIT 93] LITOW B., DUMAS P., "Additive cellular automata and algebraic series", *Theoretical Computer Science*, vol. 119, pp. 345–354, 1993.

[LOR 95] LORAUD N., "$\beta$-shift, systèmes de numération et automates", *Journal de Théorie des Nombres de Bordeaux*, vol. 7, pp. 473–498, 1995.

[LOT 83] LOTHAIRE M., *Combinatorics on Words*, Encyclopedia of Mathematics and Its Applications, Addison-Wesley, vol. 17, 1983.

[LOT 02] LOTHAIRE M., *Algebraic Combinatorics on Words*, Encyclopedia of Mathematics and Its Applications, Cambridge University Press, vol. 90, 2002.

[LOT 05] LOTHAIRE M., *Applied Combinatorics on Words*, Encyclopedia of Mathematics and Its Applications, Cambridge University Press, vol. 105, 2005.

[LUC 88] DE LUCA A., VARRICCHIO S., "On the factors of the Thue-Morse word on three symbols", *Information Processing Letters*, vol. 27, pp. 281–285, 1988.

[LUC 99] DE LUCA A., VARRICCHIO S., *Finiteness and Regularity in Semigroups and Formal Languages*, Monographs in Theoretical Computer Science, EATCS Series, Springer-Verlag, Berlin, 1999.

[MAD 13] MADILL B., RAMPERSAD N., "The abelian complexity of the paperfolding word", *Discrete Mathematics*, vol. 313, no. 7, pp. 831–838, 2013.

[MAE 99] MAES A., "An automata-theoretic decidability proof for first-order theory of $\langle \mathbb{N}, <, P \rangle$ with morphic predicate $P$", *Journal of Automata, Languages, and Combinatorics*, vol. 4, pp. 229–245, 1999.

[MAN 12] MANCE B., "Cantor series constructions of sets of normal numbers", *Acta Arithmetica*, vol. 156, no. 3, pp. 223–245, 2012.

[MAR 82] MARKOFF A.A., "Sur une question de Jean Bernoulli", *Mathematische Annalen*, vol. 19, pp. 27–36, 1882.

[MAR 13] MARSAULT V., SAKAROVITCH J., "Ultimate periodicity of $b$-recognisable sets: a quasilinear procedure", *DLT, Marne-la-Vallée*, Lecture Notes in Computer Science, Springer-Verlag, vol. 7907, pp. 362–373, 2013.

[MAS 04] MASÁKOVÁ Z., "Propriétés arithmétiques et combinatoires des $\beta$-entiers", *Annales des Sciences Mathématiques du Québec*, vol. 28, nos. 1–2, pp. 153–164, 2004.

[MAS 11] MASÁKOVÁ Z., PELANTOVÁ E., VÁVRA T., "Arithmetics in number systems with a negative base", *Theoretical Computer Science*, vol. 412, nos. 8–10, pp. 835–845, 2011.

[MAS 13] MASÁKOVÁ Z., PELANTOVÁ E., "Purely periodic expansions in systems with negative base", *Acta Mathematica Hungarica*, vol. 139, no. 3, pp. 208–227, 2013.

[MAT 05] MATIYASEVICH Y., SÉNIZERGUES G., "Decision problems for semi-Thue systems with a few rules", *Theoretical Computer Science*, vol. 330, no. 1, pp. 145–169, 2005.

[MAT 12] MATOMÄKI K., SAARI K., "A new geometric approach to Sturmian words", *Theoretical Computer Science*, vol. 432, pp. 77–84, 2012.

[MIC 96] MICHAUX C., VILLEMAIRE R., "Presburger arithmetic and recognizability of sets of natural numbers by automata: new proofs of Cobham's and Semenov's theorems", *Annals of Pure and Applied Logic*, vol. 77, pp. 251–277, 1996.

[MIG 91] MIGNOSI F., "On the number of factors of Sturmian words", *Theoretical Computer Science*, vol. 82, pp. 71–84, 1991.

[MIG 92] MIGNOSI F., PIRILLO G., "Repetitions in the Fibonacci infinite word", *RAIRO Informatique Théorique et Applications*, vol. 26, pp. 199–204, 1992.

[MIG 93] MIGNOSI F., SÉÉBOLD P., "Morphismes sturmiens et règles de Rauzy", *Journal de Théorie des Nombres de Bordeaux*, vol. 5, pp. 221–233, 1993.

[MIT 11] MITROFANOV I., "A proof for the decidability of HD0L ultimate periodicity", 2011. Available at http://arxiv.org/abs/1110.4780

[MOR 21] MORSE M., "Recurrent geodesics on a surface of negative curvature", *Transactions of the American Mathematical Society*, vol. 22, pp. 84–100, 1921.

[MOR 38] MORSE M., HEDLUND G.A., "Symbolic dynamics", *American Journal of Mathematics*, vol. 60, pp. 815–866, 1938.

[MOR 40] MORSE M., HEDLUND G.A., "Symbolic dynamics II. Sturmian trajectories", *American Journal of Mathematics*, vol. 62, pp. 1–42, 1940.

[MOS 92] MOSSÉ B., "Puissances de mots et reconnaissabilité des points fixes d'une substitution", *Theoretical Computer Science*, vol. 99, pp. 327–334, 1992.

[MOS 09] MOSHE Y., "On the joint subword complexity of automatic sequences", *Theoretical Computer Science*, vol. 410, no. 38–40, pp. 3573–3588, 2009.

[MOU 92] MOULIN-OLLAGNIER J., "Proof of Dejean's conjecture for alphabets with $5,6,7,8,9,10$ and $11$ letters", *Theoretical Computer Science*, vol. 95, pp. 187–205, 1992.

[NAK 12] NAKANO F., SADAHIRO T., "A $(-\beta)$-expansion associated to Sturmian sequences", *Integers*, vol. 12, no. 4, pp. 571–599, 2012.

[NIC 07] NICOLAY S., RIGO M., "About frequencies of letters in generalized automatic sequences", *Theoretical Computer Science*, vol. 374, nos. 1–3, pp. 25–40, 2007.

[NOU 05] NOUVEL B., RÉMILA É., "Configurations induced by discrete rotations: periodicity and quasi-periodicity properties", *Discrete Applied Mathematics*, vol. 147, nos. 2–3, pp. 325–343, 2005.

[OCH 12] OCHEM P., VASLET E., "Repetition thresholds for subdivided graphs and trees", *RAIRO Theoretical Informatics and Applications*, vol. 46, no. 1, pp. 123–130, 2012.

[OLD 39] OLDENBURGER R., "Exponent trajectories in symbolic dynamics", *Transactions of the American Mathematical Society*, vol. 46, pp. 453–466, 1939.

[OPP 78] OPPEN D.C., "A superexponential upper bound on the complexity of Presburger arithmetic", *Journal of Computer and System Sciences*, vol. 16, no. 3, pp. 323–332, 1978.

[OST 22] OSTROWSKI A., "Bemerkungen zur Theorie der Diophantischen Approximationen", *Abhandlungen aus dem Mathematischen Seminar der Universitat Hamburg*, vol. 1, pp. 77–98, 250–251, 1922.

[PAN 83] PANSIOT J.-J., "Hiérarchie et fermeture de certaines classes de tag-systèmes", *Acta Informatica*, vol. 20, pp. 179–196, 1983.

[PAN 84a] PANSIOT J.-J., "Complexité des facteurs des mots infinis engendrés par morphismes itérés", PAREDAENS J., (ed.), *Proceedings of the 11th International Conference on Automata, Languages, and Programming (ICALP)*, Lecture Notes in Computer Science, Springer-Verlag, vol. 172, pp. 380–389, 1984.

[PAN 84b] PANSIOT J.-J., "A propos d'une conjecture de F. Dejean sur les répétitions dans les mots", *Discrete Applied Mathematics*, vol. 7, pp. 297–311, 1984.

[PAN 85a] PANSIOT J.-J., "On various classes of infinite words obtained by iterated mappings", in NIVAT M., PERRIN D., (eds.), *Automata on Infinite Words*, Lecture Notes in Computer Science, Springer-Verlag, vol. 192, pp. 188–197, 1985.

[PAN 85b] PANSIOT J.-J., "Subword complexities and iteration", *Bulletin European Association for Theoretical Computer Science*, vol. 26, pp. 55–62, 1985.

[PAN 86] PANSIOT J.-J., "Decidability of periodicity for infinite words", *RAIRO Theoretical Informatics and Applications*, vol. 20, pp. 43–46, 1986.

[PAP 94] PAPADIMITRIOU C.H., *Computational Complexity*, Addison-Wesley Publishing Company, Reading, MA, 1994.

[PAR 60] PARRY W., "On the $\beta$-expansions of real numbers", *Acta Mathematica Academiae Scientiarum Hungaricae*, vol. 11, pp. 401–416, 1960.

[PAR 66] PARIKH R.J., "On context-free languages", *Journal of Association for Computing Machinery*, vol. 13, pp. 570–581, 1966.

[PAR 14] PARREAU A., ROWLAND E., RIGO M., VANDOMME É., "A new approach to the 2-regularity of the $\ell$-abelian complexity of 2-automatic sequences", 2014. Available at http://arxiv.org/abs/1405.3532

[PAT 70] PATERSON M.S., "Unsolvability in $3 \times 3$ matrices", *Studies in Applied Mathematics*, vol. 49, pp. 105–107, 1970.

[PĂU 95] PĂUN G., SALOMAA A., "Thin and slender languages", *Discrete Applied Mathematics*, vol. 61, no. 3, pp. 257–270, 1995.

[PEN 65] PENNEY W., "A "binary" system for complex numbers", *Journal of Association for Computing Machinery,* vol. 12, pp. 247–248, 1965.

[PER 90] PERRIN D., "Finite automata", in VAN LEEUWEN J., (ed.), *Handbook of Theoretical Computer Science, Volume B: Formal Models and Semantics,* Elsevier – MIT Press, pp. 1–57, 1990.

[PER 95] PERRIN D., "Les débuts de la théorie des automates", *Technique et Science Informatique,* vol. 14, pp. 409–443, 1995.

[PER 04] PERRIN D., PIN J.-É., *Infinite Words. Automata, Semigroups, Logic and Games,* Amsterdam: Elsevier/Academic Press, 2004.

[PET 91] PETHÖ A., "On a polynomial transformation and its application to the construction of a public key cryptosystem", in *Computational Number Theory (Debrecen, 1989),* de Gruyter, Berlin, pp. 31–43, 1991.

[PEY 87] PEYRIÈRE J., "Fréquence des motifs dans les suites doubles invariantes par une substitution", *Annales des Sciences Mathématiques du Québec,* vol. 11, pp. 133–138, 1987.

[PIS 46] PISOT C., "Répartition (mod1) des puissances successives des nombres réels", *Commentarii Mathematici Helvetici,* vol. 19, pp. 153–160, 1946.

[POI 97] POINT F., BRUYÈRE V., "On the Cobham-Semenov theorem," *Theory of Computing Systems,* vol. 30, pp. 197–220, 1997.

[PRE 91] PRESBURGER M., "On the completeness of a certain system of arithmetic of whole numbers in which addition occurs as the only operation", *History and Philosophy of Logic,* vol. 12, no. 2, pp. 225–233, 1991.

[PRO 51] PROUHET E., "Mémoire sur quelques relations entre les puissances des nombres", *Comptes rendus de l'Académie des Sciences Paris,* vol. 33, p. 225, 1851.

[PRO 79] PRODINGER H., URBANEK F.J., "Infinite 0–1-sequences without long adjacent identical blocks", *Discrete mathematics,* vol. 28, pp. 277–289, 1979.

[PUZ 13] PUZYNINA S., ZAMBONI L.Q., "Abelian returns in Sturmian words", *Journal of Combinatorial Theory, Series A,* vol. 120, no. 2, pp. 390–408, 2013.

[PYT 02] PYTHEAS FOGG N., *Substitutions in Dynamics, Arithmetics and Combinatorics*, in BERTHÉ V., FERENCZI S., MAUDUIT C., SIEGEL A., (eds.), Lecture Notes in Mathematics, vol. 1794, Springer-Verlag, 2002.

[QUA 04] QUAS A., ZAMBONI L., "Periodicity and local complexity", *Theoretical Computer Science*, vol. 319, nos. 1–3, pp. 229–240, 2004.

[QUE 87] QUEFFÉLEC M., *Substitution Dynamical Systems – Spectral Analysis*, Lecture Notes in Mathematics, vol. 1294, Springer-Verlag, 1987.

[RAM 05] RAMPERSAD N., SHALLIT J., WANG M.-W., "Avoiding large squares in infinite binary words", *Theoretical Computer Science*, vol. 339, no. 1, pp. 19–34, 2005.

[RAM 11] RAMPERSAD N., "Abstract numeration systems", In *Language and Automata Theory and Applications. 5th International Conference, LATA, Tarragona. Proceedings*, Lecture Notes in Computer Science, Springer-Verlag, vol. 6638, pp. 65–79, 2011.

[RAO 13] RAO M., RIGO M., SALIMOV P., "Avoiding 2-binomial squares and cubes", 2013. Available at http://arxiv.org/abs/1310.4743 .

[RAU 82] RAUZY G., "Nombres algébriques et substitutions", *Bulletin de la Société Mathématique de France*, vol. 110, pp. 147–178, 1982.

[RAV 88] RAVENSTEIN T.V., "The three gap theorem (Steinhaus conjecture)", *Journal of the Australian Mathematical Society Series A*, vol. 45, pp. 360–370, 1988.

[RÉN 57] RÉNYI A., "Representations for real numbers and their ergodic properties", *Acta Mathematica Academiae Scientiarum Hungaricae*, vol. 8, pp. 477–493, 1957.

[RES 12] RESTIVO A., ROSONE G., "On the product of balanced sequences", *RAIRO Theoretical Informatics and Applications*, vol. 46, no. 1, pp. 131–145, 2012.

[RIC 99] RICHOMME G., "Test-words for Sturmian morphisms", *Bulletin of the Belgian Mathematical Society Simon Stevin*, vol. 6, no. 4, 481–489.

[RIC 10] RICHOMME G., SAARI K., ZAMBONI L.Q., "Balance and abelian complexity of the Tribonacci word", *Advances in Applied Mathematics*, vol. 45, no. 2, pp. 212–231, 2010.

[RIC 11] RICHOMME G., SAARI K., ZAMBONI L.Q., "Abelian complexity of minimal subshifts", *Journal of the London Mathematical Society (2)*, vol. 83, no. 1, pp. 79–95, 2011.

[RIG 00] RIGO M., "Generalization of automatic sequences for numeration systems on a regular language", *Theoretical Computer Science*, vol. 244, pp. 271–281, 2000.

[RIG 01] RIGO M., "Numeration systems on a regular language: arithmetic operations, recognizability and formal power series", *Theoretical Computer Science*, vol. 269, pp. 469–498, 2001.

[RIG 02a] RIGO M., "Construction of regular languages and recognizability of polynomials", *Discrete Mathematics*, vol. 254, pp. 485–496, 2002.

[RIG 02b] RIGO M., MAES A., "More on generalized automatic sequences", *Journal of Automata Languages and Combinatorics*, vol. 7, no. 3, pp. 351–376, 2002.

[RIG 06] RIGO M., WAXWEILER L., "A note on syndeticity, recognizable sets and Cobham's theorem", *Bulletin of the European Association for Theoretical Computer Science EATCS*, vol. 88, pp.169–173, 2006.

[RIG 10] RIGO M., "Numeration systems: a link between number theory and formal language theory", In *Developments in language theory*, Lecture Notes in Computer Science, Springer-Verlag, Berlin, vol. 6224, pp. 33–53, 2010.

[RIG 13a] RIGO M., SALIMOV P., VANDOMME E., "Some properties of abelian return words", *Journal of Integer Sequences*, vol. 16, no. 2, Article 13.2.5., 2013

[RIG 13b] RIGO M., SALIMOV P., "Another generalization of abelian equivalence: binomial complexity of infinite words", In *WORDS (Turku, Finland)*, Lecture Notes in Computer Science, Springer-Verlag, Berlin, vol. 8079, pp. 217–228, 2013.

[RIG 14] RIGO M., *Formal Languages, Automata and Numeration Systems 1: Introduction to Combinations on Words*, ISTE, London and John Wiley & Sons, New York, 2014.

[ROT 94] ROTE G., "Sequences with subword complexity $2n$", *Journal of Number Theory*, vol. 46, pp. 196–213, 1994.

[ROW 12a] ROWLAND E., YASSAWI R., "A characterization of p-automatic sequences as columns of linear cellular automata", 2012. Available at http://arxiv.org/abs/1209.6008

[ROW 12b] ROWLAND E., SHALLIT J., "$k$-automatic sets of rational numbers", *Language and Automata Theory and Applications*, Lecture Notes in Computer Science, Springer-Verlag, vol. 7183, pp. 490–501, 2012.

[ROZ 97] ROZENBERG G., SALOMAA A., (eds.), *Handbook of Formal Languages. Word, Language, Grammar*, Springer-Verlag, Berlin, vol. 1, 1997.

[RYO 13] RYOMA S., "Text compression using abstract numeration system on a regular language", 2013. Available at http://arxiv.org/abs/1308.0267

[SAA 06] SAARI K., "On the frequency of letters in morphic sequences", *Computer Science—Theory and Applications*, Lecture Notes in Computer Science, Springer-Verlag, Berlin, vol. 3967, pp. 334–345, 2006.

[SAK 09] SAKAROVITCH J., *Elements of Automata Theory*, Cambridge University Press, Cambridge, 2009.

[SAL 87a] SALON O., "Suites automatiques à multi-indices", In *Séminaire de Théorie des Nombres de Bordeaux*, pp. 4.01–4.27, 1986–1987.

[SAL 87b] SALON O., "Suites automatiques à multi-indices et algébricité", *Comptes Rendus de l'Académie des Sciences Paris*, vol. 305, pp. 501–504, 1987.

[SAL 10] SALIMOV P.V., "On uniform recurrence of a direct product", *Discrete Mathematics & Theoretical Computer Science*, vol. 12, no. 4, pp. 1–8, 2010.

[SCH 80] SCHMIDT K., "On periodic expansions of Pisot numbers and Salem numbers", *Bulletin of the London Mathematical Society*, vol. 12, no. 4, pp. 269–278, 1980.

[SCH 12] SCHAEFFER L., SHALLIT J., "The critical exponent is computable for automatic sequences", *International Journal of Foundations of Computer Science*, vol. 23, no. 8, pp. 1611–1626, 2012.

[SEN 81] SENETA E., *Nonnegative Matrices and Markov Chains*, 2nd ed., Springer-Verlag, New York, 1981 .

[SHA 94] SHALLIT J.O., "Numeration systems, linear recurrences, and regular sets", *Information and Computation*, vol. 113, pp. 331–347, 1994.

[SHA 08] SHALLIT J., *A Second Course in Formal Languages and Automata Theory*, Cambridge University Press, Cambridge, 2008.

[SHU 08] SHUR A.M., "Combinatorial complexity of regular languages", *Computer Science—Theory and Applications*, Lecture Notes in Computer Science, Springer-Verlag, Berlin, vol. 5010, pp. 289–301, 2008.

[SHU 10] SHUR A.M., "Growth rates of complexity of power-free languages", *Theoretical Computer Science*, vol. 411, nos. 34–36, pp. 3209–3223, 2010.

[SIE 09] SIEGEL A., THUSWALDNER J.M., "Topological properties of Rauzy fractals", *Mémoires de la Société Mathématique de France*, vol. 118, p.140, 2009.

[STA 82] STANAT D., WEISS S.F., "A pumping theorem for regular languages", *ACM SIGACT News*, vol. 14, no.1, pp. 36–37, 1982.

[STE 13] STEINER W., "Digital expansions with negative real bases", *Acta Mathematica Hungarica*, vol. 139, nos. 1–2, pp. 106–119, 2013.

[STR 94] STRAUBING H., *Finite automata, formal logic, and circuit complexity*, Progress in Theoretical Computer Science, Birkhäuser Boston, Inc., Boston, MA, 1994.

[SUD 06] SUDKAMP T.A., *Languages and Machines: An Introduction to the Theory of Computer Science*, Addison-Wesley, 2006.

[SZI 92] SZILARD A., YU S., ZHANG K., SHALLIT J., "Characterizing regular languages with polynomial densities", *Mathematical Foundations of Computer Science 1992 (Prague, 1992)*, Vol. 629 Lecture Notes in Computer Science, Springer-Verlag, Berlin, pp. 494–503, 1992.

[TAN 07] TAN B., WEN Z.-Y., "Some properties of the Tribonacci sequence", *European Journal of Combinatorics*, vol. 28, no. 6, pp. 1703–1719, 2007.

[THO 90] THOMAS W., "Automata on infinite objects", in VAN LEEUWEN J., (ed.), *Handbook of Theoretical Computer Science, Volume B: Formal Models and Semantics,* Elsevier – MIT Press, pp. 133–191, 1990.

[THU 12] THUE A., "Über die gegenseitige Lage gleicher Teile gewisser Zeichenreihen", *Norske vid. Selsk. Skr. Mat. Nat. Kl.,* vol. 1, pp. 1–67, 1912.

[THU 89] THURSTON W.P., "Groups, tilings, and finite state automata, AMS Colloquium Lecture Notes", American Mathematical Society, 1989.

[TIJ 06] TIJDEMAN R., "Periodicity and almost-periodicity], in *More Sets, Graphs and Numbers, Bolyai Soc. Math. Stud.,* Springer-Verlag, Berlin, vol. 15, pp. 381–405, 2006.

[TOE 28] TOEPLITZ O., "Beispiele zur Theorie der fastperiodischen Funktionen", *Mathematische Annalen,* vol. 98, pp. 281–295, 1928.

[TUR 13a] TUREK O., "Abelian complexity function of the tribonacci word", 2013. Available at http://arxiv.org/abs/1309.4810

[TUR 13b] TUREK O., "Abelian complexity and abelian co-decomposition", *Theoretical Computer Science,* vol. 469, pp. 77–91, 2013.

[VAR 91] VARDI I., *Computational Recreations in Mathematica,* Addison-Wesley Publishing Company, Advanced Book Program, Redwood City, CA, 1991.

[VEN 70] VENKOV B.A., *Elementary Number Theory,* Wolters-Noordhoff, Groningen, 1970.

[VIL 92] VILLEMAIRE R., "The theory of $\langle \mathbf{N}, +, V_k, V_l \rangle$ is undecidable", *Theoretical Computer Science,* vol. 106, pp. 337–349, 1992.

[VOL 08] VOLKOV M.V., "Synchronizing automata and the Cerný conjecture", *Language and Automata Theory and Applications,* Lecture Notes in Computer Science, Springer-Verlag, Berlin, vol. 5196, pp. 11–27, 2008.

[WAR 08] WARD R., "On robustness properties of beta encoders and golden ratio encoders", *IEEE Transactions on Information Theory,* vol. 54, no. 9, pp. 4324–4334, 2008.

[WEY 16] WEYL H., "Über die Gleichverteilung von Zahlen mod. Eins", *Mathematische Annalen,* vol. 77, no. 3, pp. 313–352, 1916.

[WIC 10] WICKERHAUSER M.V., *Mathematics for multimedia*, Applied and Numerical Harmonic Analysis, Birkhäuser Boston Inc., Boston, MA, 2010, Corrected reprint of the 2004 original.

[WIL 04] WILLARD S., *General Topology*, Dover Publications Inc., Mineola, NY, 2004.

[WOL 02] WOLFRAM S., *A New Kind of Science*, Wolfram Media, 2002.

[WOO 78] WOODS D.R., "Elementary problem proposal E 2692", *American Mathematical Monthly*, vol. 85, p. 48, 1978.

[WOO 79] WOODS D.R., ROBBINS D., GRIPENBERG G., "Problems and solutions: solutions of elementary problems: E2692", *American Mathematical Monthly*, vol. 86, no. 5, pp. 394–395, 1979.

[WYT 07] WYTHOFF W.A., "A modification of the game of nim", *Nieuw Archief voor Wiskunde*, vol. 7, pp. 199–202, 1907.

[YAS 99] YASUTOMI S.-I., "On Sturmian sequences which are invariant under some substitutions", in KANEMITSU S., GYÖRY K., (eds.), *Number Theory and Its Applications*, Kluwer, pp. 347–373, 1999.

[YU 97] YU S., "Regular languages", in ROZENBERG G., SALOMAA A., (eds.), *Handbook of Formal Languages*, Springer-Verlag, vol. 1, pp. 41–110, 1997.

[YU 05] YU S., "State complexity: recent results and open problems", *Fundamenta Informaticae*, vol. 64, nos. 1–4, pp. 471–480, 2005.

[ZEC 72] ZECKENDORF E., "Représentation des nombres naturels par une somme de nombres de Fibonacci ou de nombres Lucas", *Bulletin de la Société Royale des Sciences de Liège*, vol. 41, pp. 179–182, 1972.

[ZEH 10] ZEHNDER E., *Lectures on dynamical systems*, EMS Textbooks in Mathematics, European Mathematical Society (EMS), Zürich, 2010, Hamiltonian vector fields and symplectic capacities.

# Index

# Volume 1 - Contents

# Volume 1 - Index

Other titles from

in

Computer Engineering

## 2014

BOULANGER Jean-Louis
*Formal Methods Applied to Industrial Complex Systems*

BOULANGER Jean-Louis
*Formal Methods Applied to Complex Systems: Implementation of the B Method*

GARDI Frédéric, BENOIST Thierry, DARLAY Julien, ESTELLON Bertrand, MEGEL Romain
*Mathematical Programming Solver based on Local Search*

OUSSALAH Mourad Chabane
*Software Architecture 1*

OUSSALAH Mourad Chabane
*Software Architecture 2*

QUESNEL Flavien
*Scheduling of Large-scale Virtualized Infrastructures: Toward Cooperative Management*

TOUATI Sid, DE DINECHIN Benoit
*Advanced Backend Optimization*

## 2013

ANDRÉ Etienne, SOULAT Romain
*The Inverse Method: Parametric Verification of Real-time Embedded Systems*

BOULANGER Jean-Louis
*Safety Management for Software-based Equipment*

DELAHAYE Daniel, PUECHMOREL Stéphane
*Modeling and Optimization of Air Traffic*

FRANCOPOULO Gil
*LMF — Lexical Markup Framework*

GHÉDIRA Khaled
*Constraint Satisfaction Problems*

ROCHANGE Christine, UHRIG Sascha, SAINRAT Pascal
*Time-Predictable Architectures*

WAHBI Mohamed
*Algorithms and Ordering Heuristics for Distributed Constraint Satisfaction Problems*

ZELM Martin *et al.*
*Enterprise Interoperability*

## 2012

ARBOLEDA Hugo, ROYER Jean-Claude
*Model-Driven and Software Product Line Engineering*

BLANCHET Gérard, DUPOUY Bertrand
*Computer Architecture*

BOULANGER Jean-Louis
*Industrial Use of Formal Methods: Formal Verification*

BOULANGER Jean-Louis
*Formal Method: Industrial Use from Model to the Code*

CALVARY Gaëlle, DELOT Thierry, SEDES Florence, TIGLI Jean-Yves
*Computer Science and Ambient Intelligence*

MAHOUT Vincent
*Assembly Language Programming: ARM Cortex-M3 2.0: Organization, Innovation and Territory*

MARLET Renaud
*Program Specialization*

SOTO Maria, SEVAUX Marc, ROSSI André, LAURENT Johann
*Memory Allocation Problems in Embedded Systems: Optimization Methods*

# 2011

BICHOT Charles-Edmond, SIARRY Patrick
*Graph Partitioning*

BOULANGER Jean-Louis
*Static Analysis of Software: The Abstract Interpretation*

CAFERRA Ricardo
*Logic for Computer Science and Artificial Intelligence*

HOMES Bernard
*Fundamentals of Software Testing*

KORDON Fabrice, HADDAD Serge, PAUTET Laurent, PETRUCCI Laure
*Distributed Systems: Design and Algorithms*

KORDON Fabrice, HADDAD Serge, PAUTET Laurent, PETRUCCI Laure
*Models and Analysis in Distributed Systems*

LORCA Xavier
*Tree-based Graph Partitioning Constraint*

TRUCHET Charlotte, ASSAYAG Gerard
*Constraint Programming in Music*

PANETTO Hervé, BOUDJLIDA Nacer
*Interoperability for Enterprise Software and Applications 2006 / IFAC-IFIP I-ESA'2006*

## 2005

GÉRARD Sébastien *et al.*
*Model Driven Engineering for Distributed Real Time Embedded Systems*

PANETTO Hervé
*Interoperability of Enterprise Software and Applications 2005*

Printed and bound by CPI Group (UK) Ltd, Croydon, CR0 4YY

27/10/2024

14580317-0003